The Giant Book of

Kakuro

1000 Hard Cross Sums Puzzles (10x10)

Vol. 7

Khalid Alzamili, Ph.D.

A Special Request

Your brief review could really help us.

Thank you for your support

August 2020

Copyright © 2020 Dr. Khalid Alzamili

ISBN: 9789922636450

Dr. Khalid Alzamili Pub

www.alzamili.com

Author Email : khalid@alzamili.com

CONTENTS

INTRODUCTION

Kakuro (also known as "Cross Sums") is a logical puzzle with simple rules and challenging solutions. The puzzle consists of a playing area of filled and empty cells similar to a crossword puzzle.

The rules of Kakuro are simple:

1. Each cell can contain numbers from 1 through 9.

2. The clues in the black cells tells the sum of the numbers next to that clue. (on the right or down).

3. The numbers in consecutive white cells must be unique. For example the total 6 you could have 1 and 5, 2 and 4 but not 3 and 3.

This Logic Puzzles book is packed with the following features:

- 1000 Hard Kakuro (10x10) Puzzles.

- Answers to every puzzle are provided.

- Each puzzle is guaranteed to have only one solution.

We hope this will be an entertaining and uplifting mental workout, enjoy The Giant Book of Kakuro.

Khalid Alzamili, Ph.D.

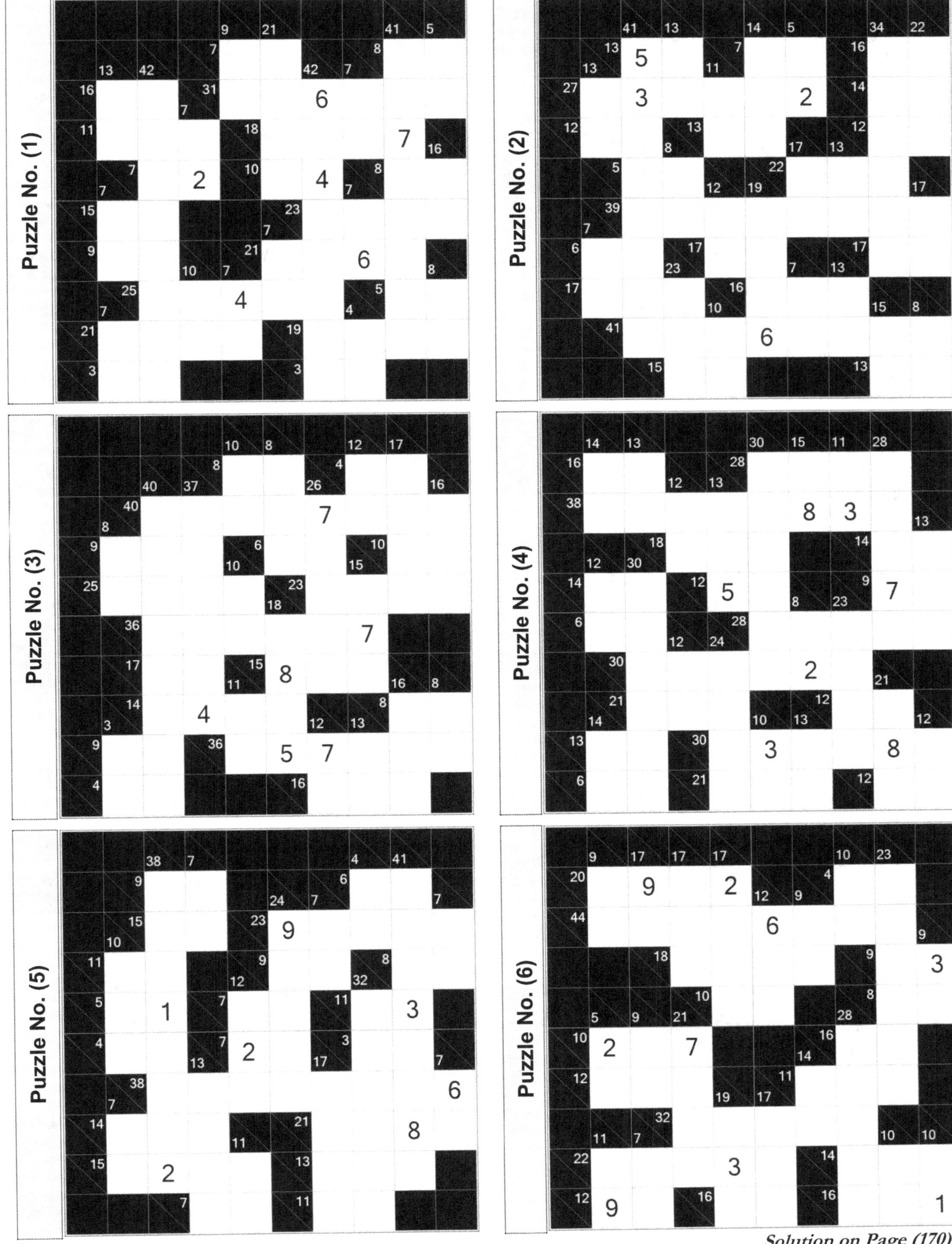

Solution on Page (170)

(3)

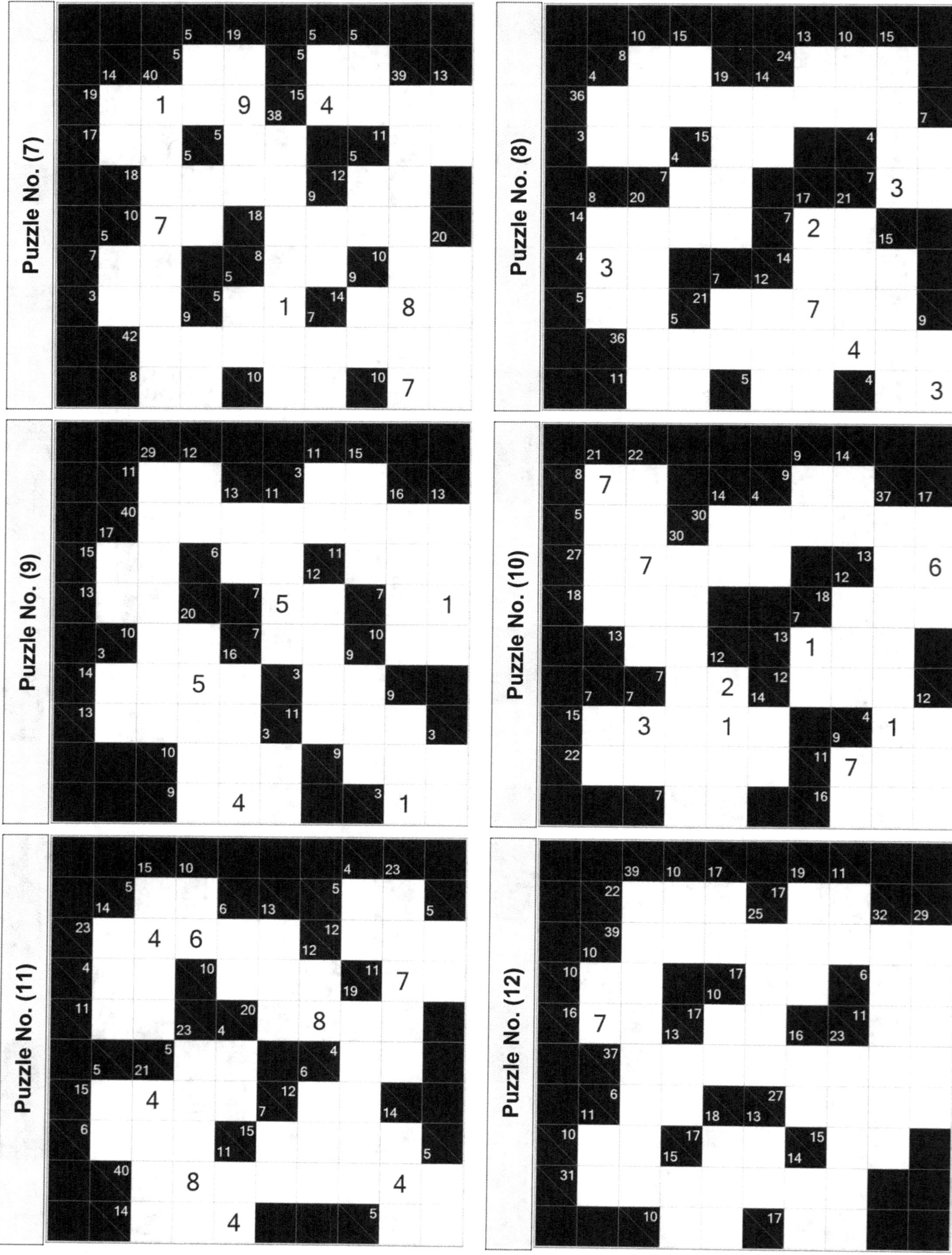

Solution on Page (170)

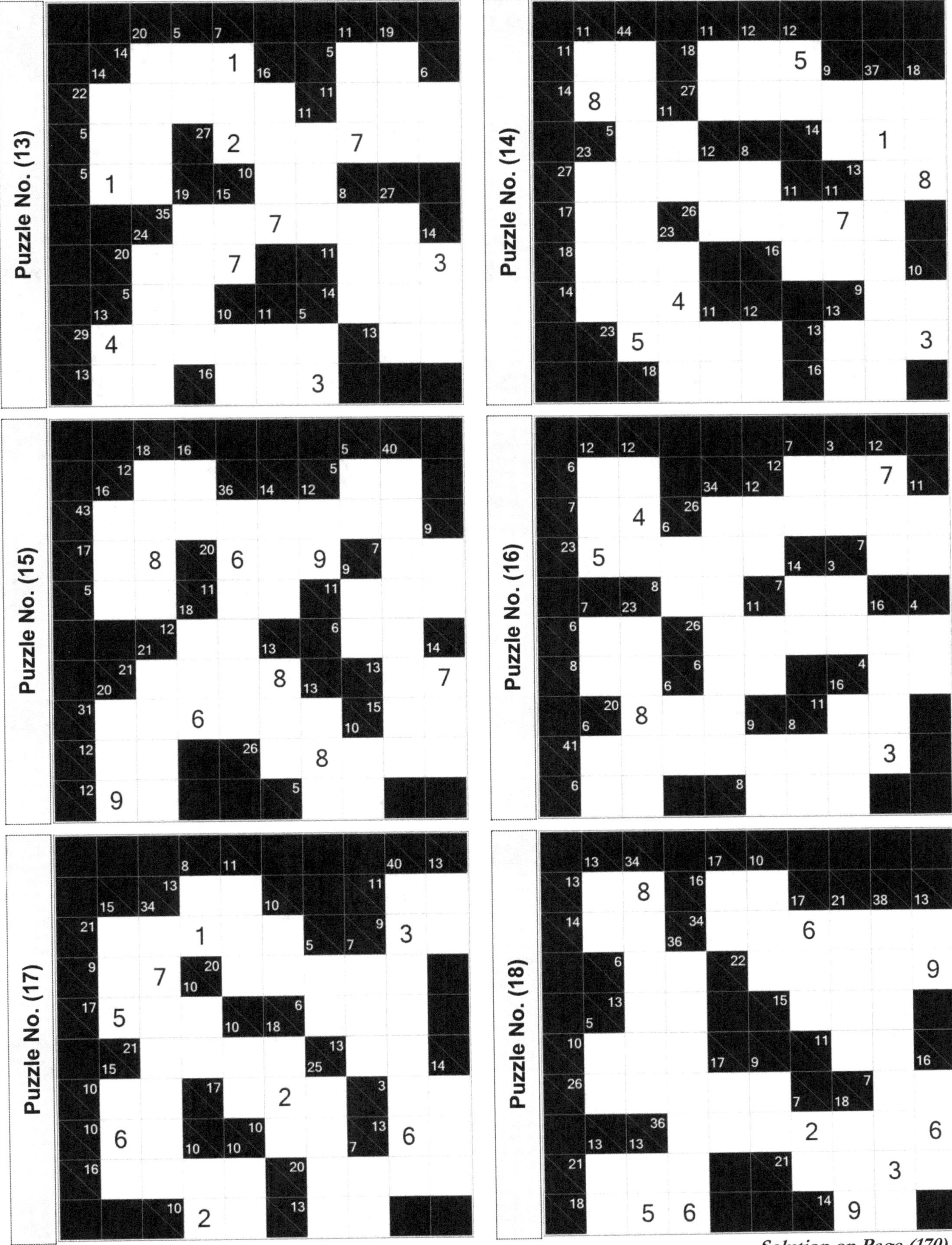

Solution on Page (170)

(5)

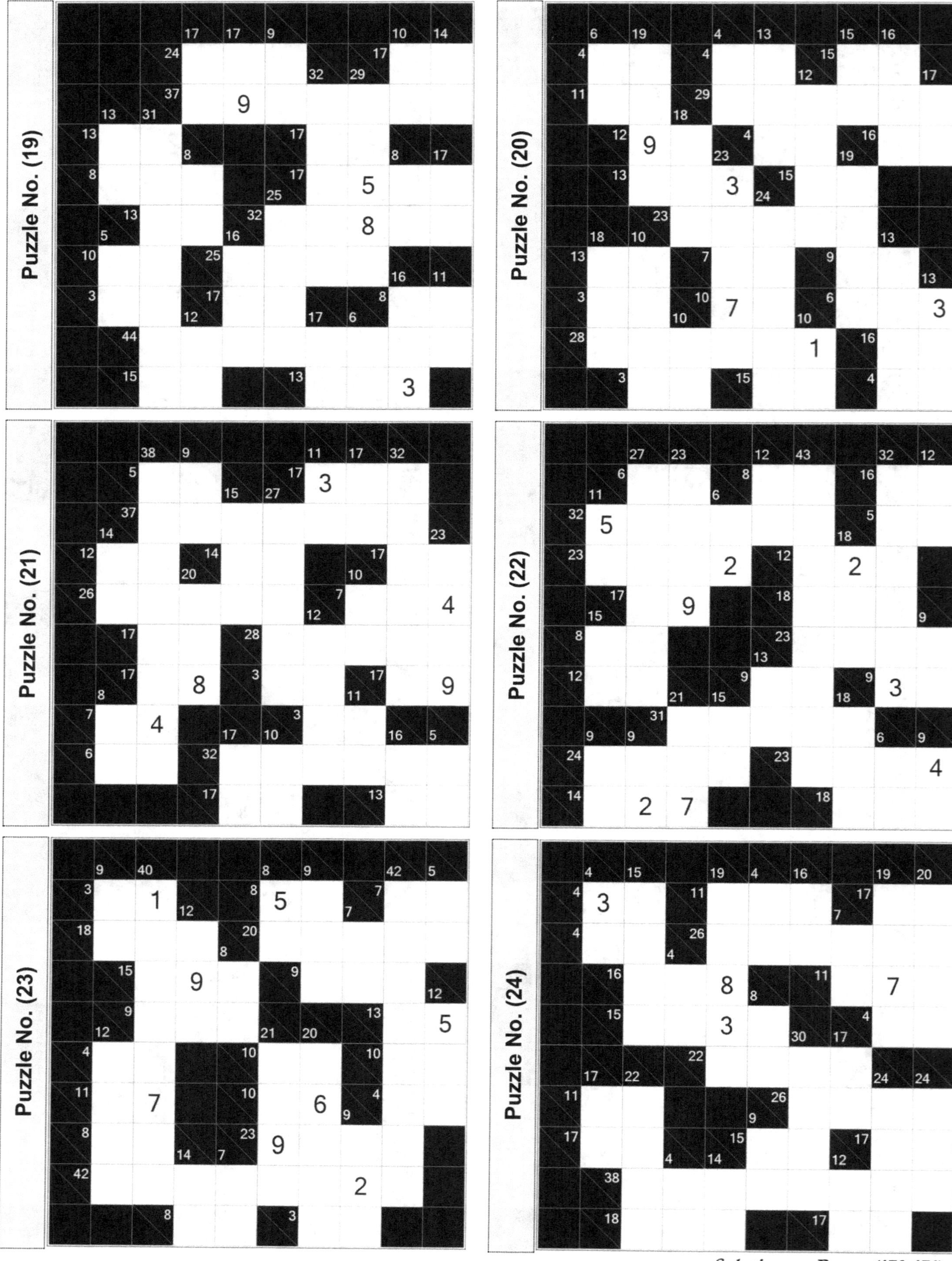

Solution on Pages (170-171)

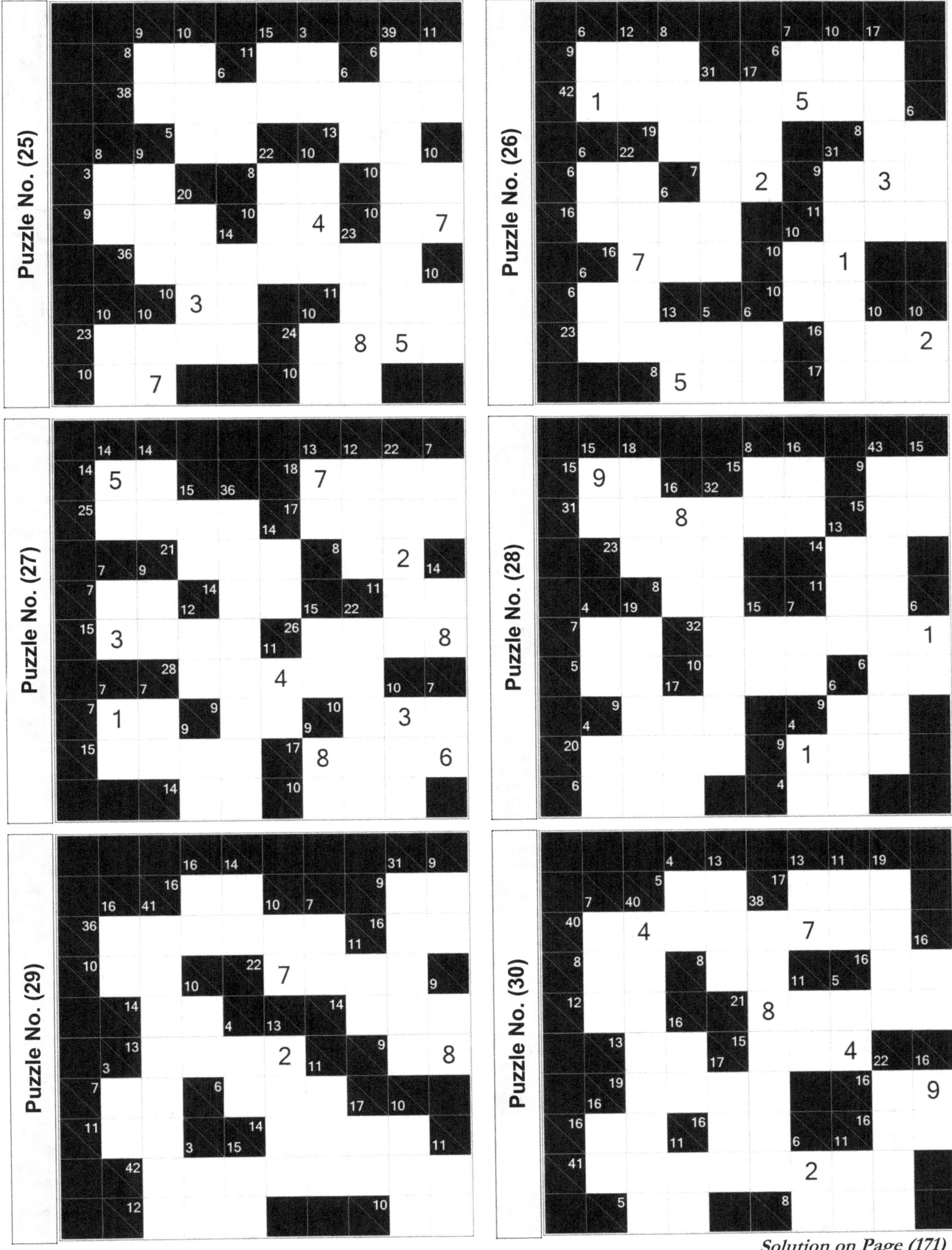

Puzzle No. (25) — Puzzle No. (26)

Puzzle No. (27) — Puzzle No. (28)

Puzzle No. (29) — Puzzle No. (30)

Solution on Page (171)

(7)

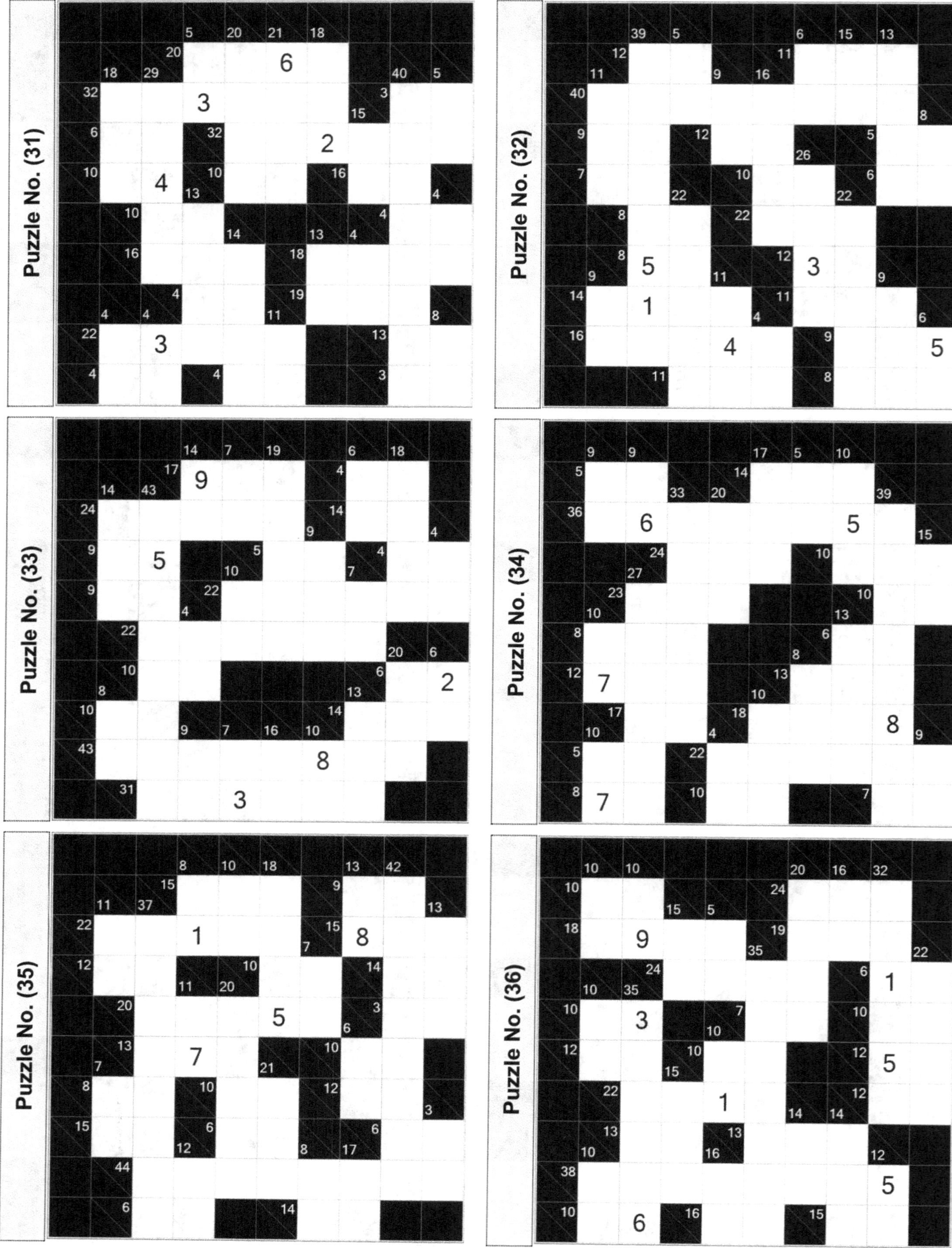

Solution on Page (171)

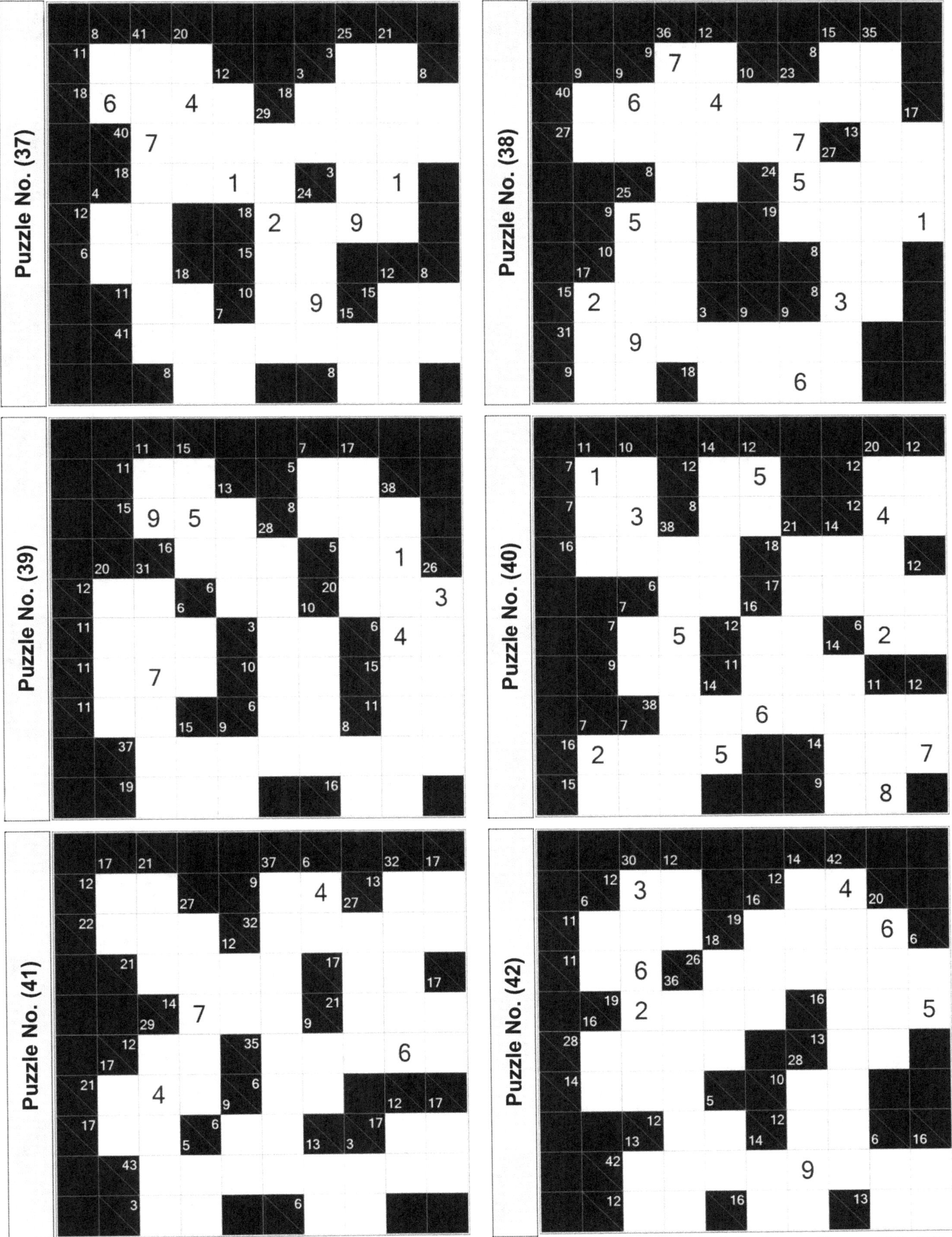

Puzzle No. (37)

Puzzle No. (38)

Puzzle No. (39)

Puzzle No. (40)

Puzzle No. (41)

Puzzle No. (42)

Solution on Pages (171-172)

(9)

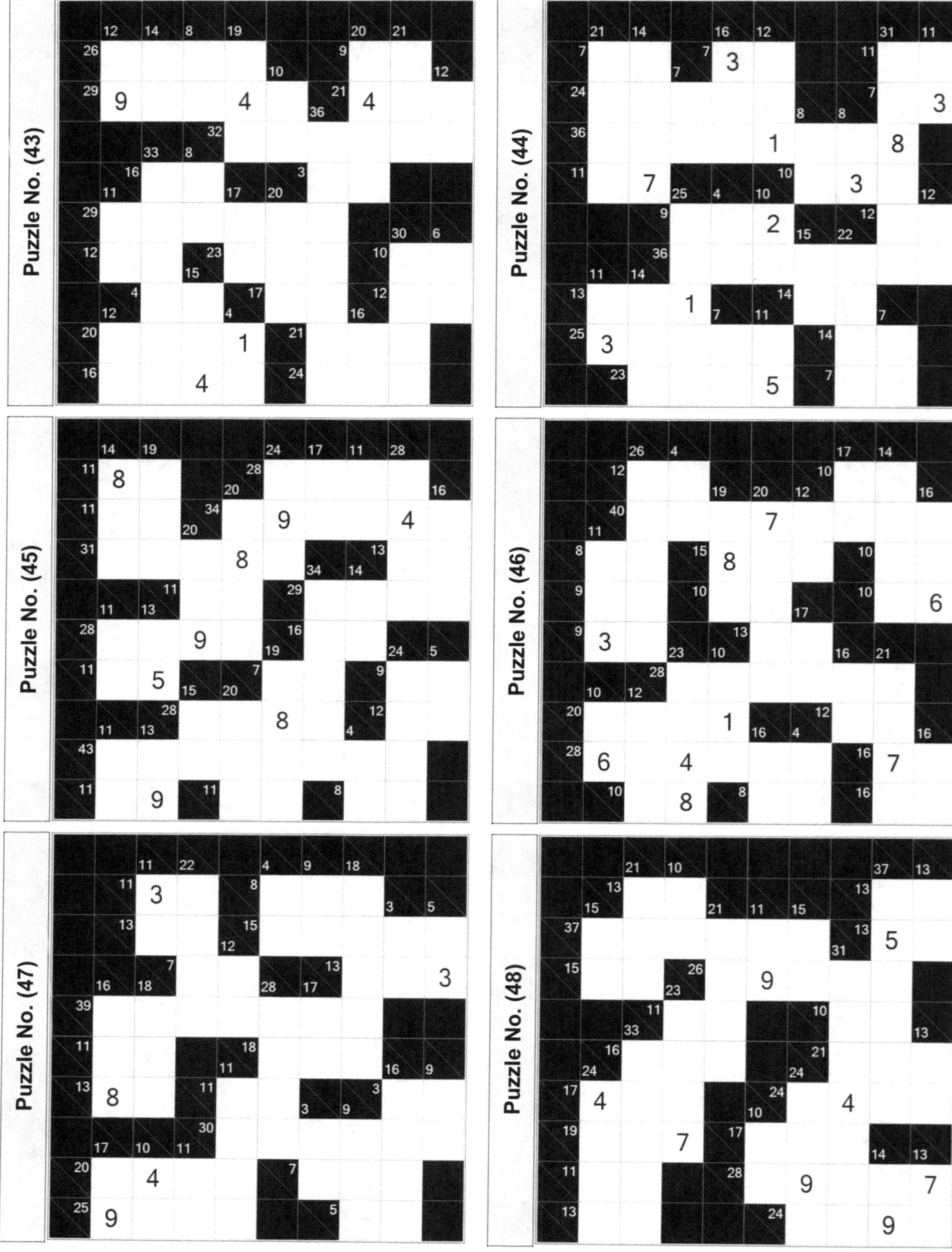

Puzzle No. (43)

Puzzle No. (44)

Puzzle No. (45)

Puzzle No. (46)

Puzzle No. (47)

Puzzle No. (48)

Solution on Page (172)

(10)

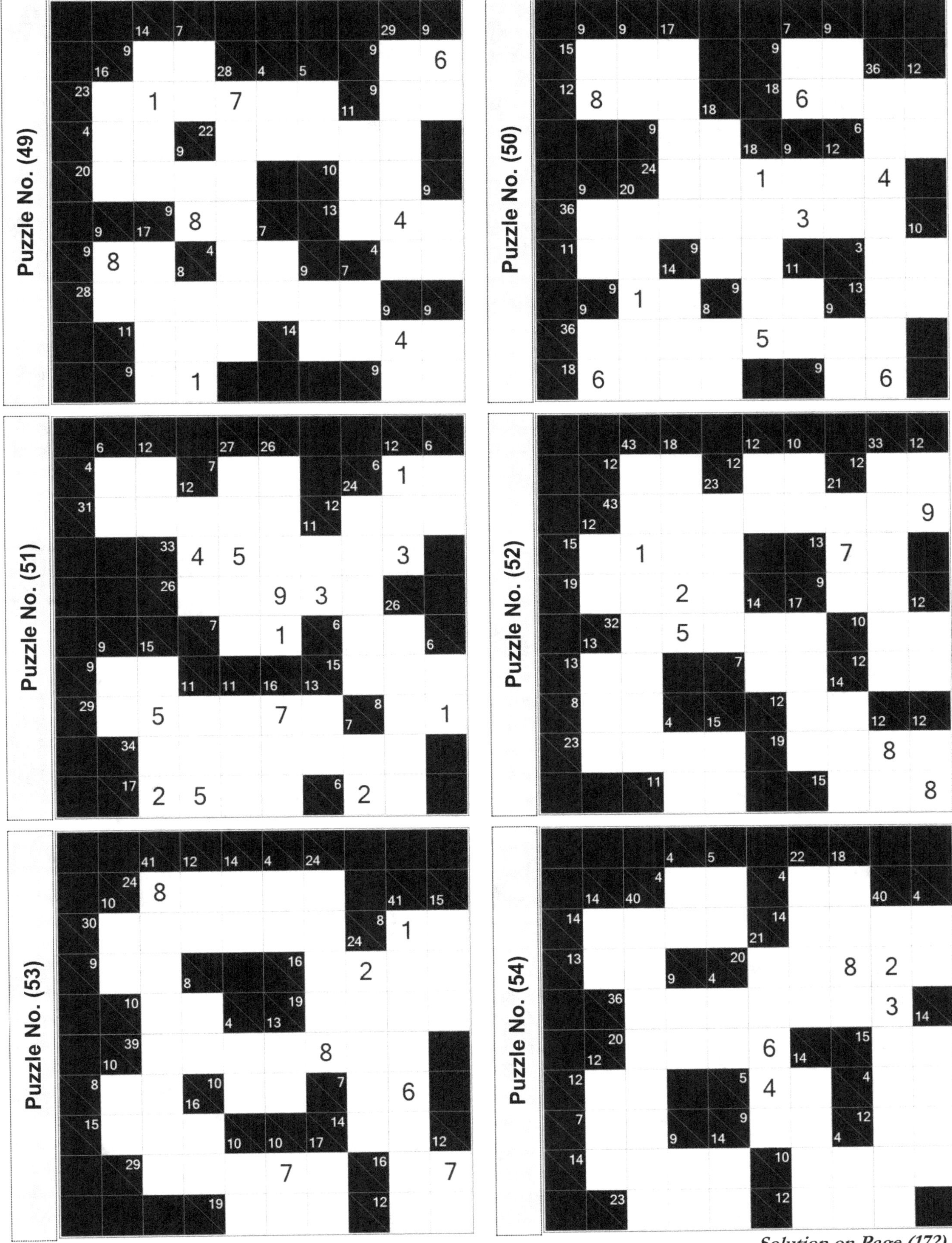

Puzzle No. (49)
Puzzle No. (50)
Puzzle No. (51)
Puzzle No. (52)
Puzzle No. (53)
Puzzle No. (54)
Solution on Page (172)

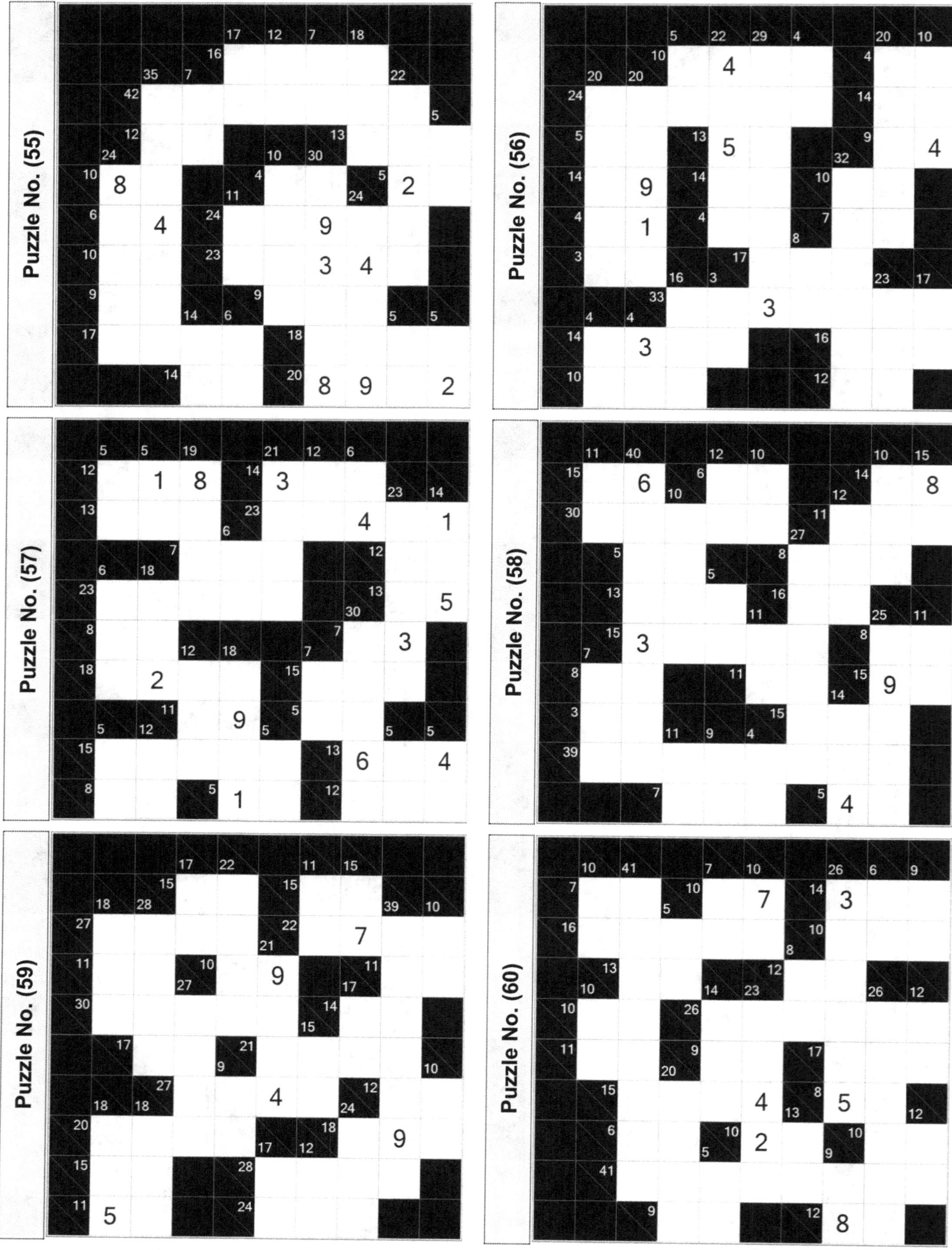

Solution on Page (172)

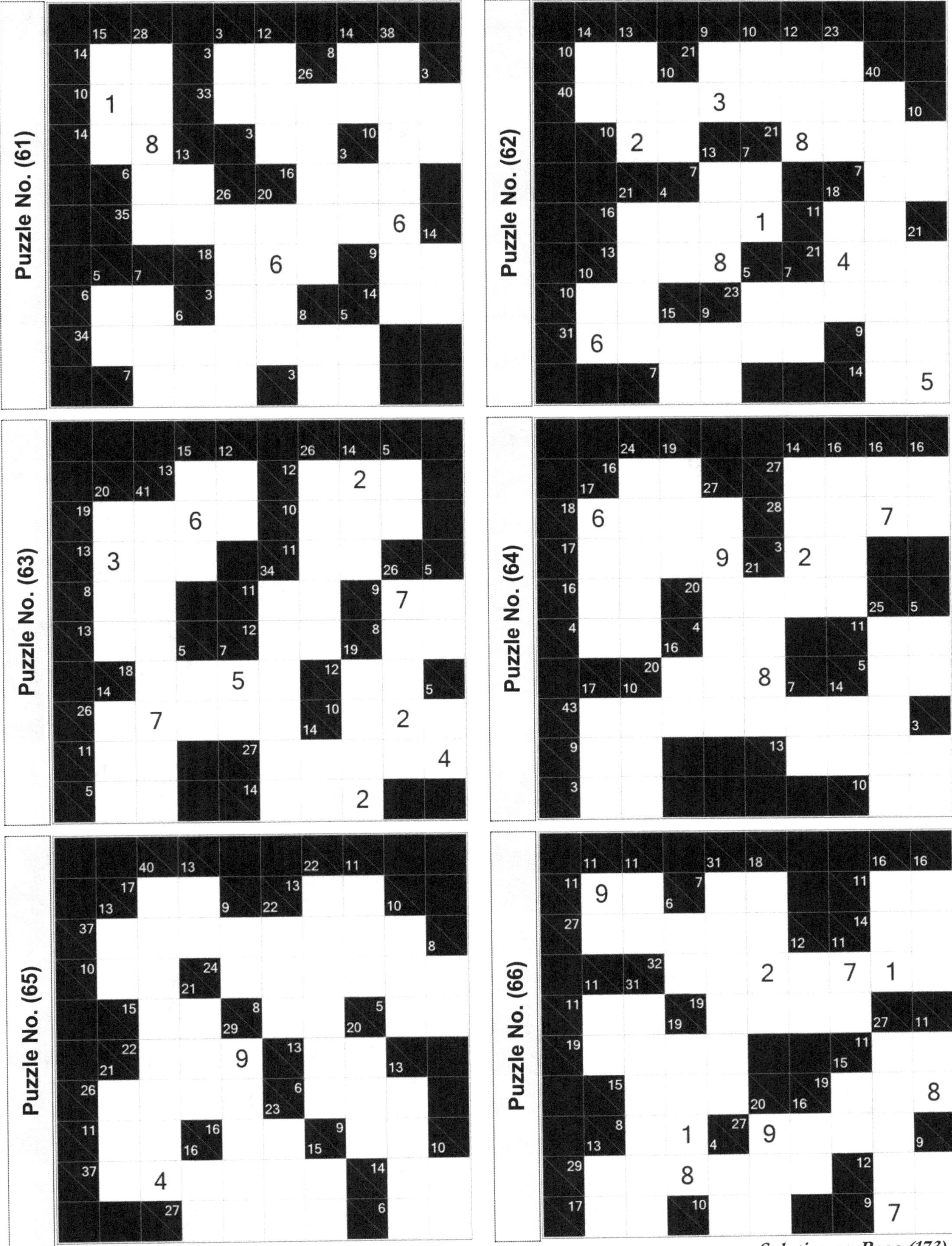

Puzzle No. (61)

Puzzle No. (62)

Puzzle No. (63)

Puzzle No. (64)

Puzzle No. (65)

Puzzle No. (66)

Solution on Page (173)

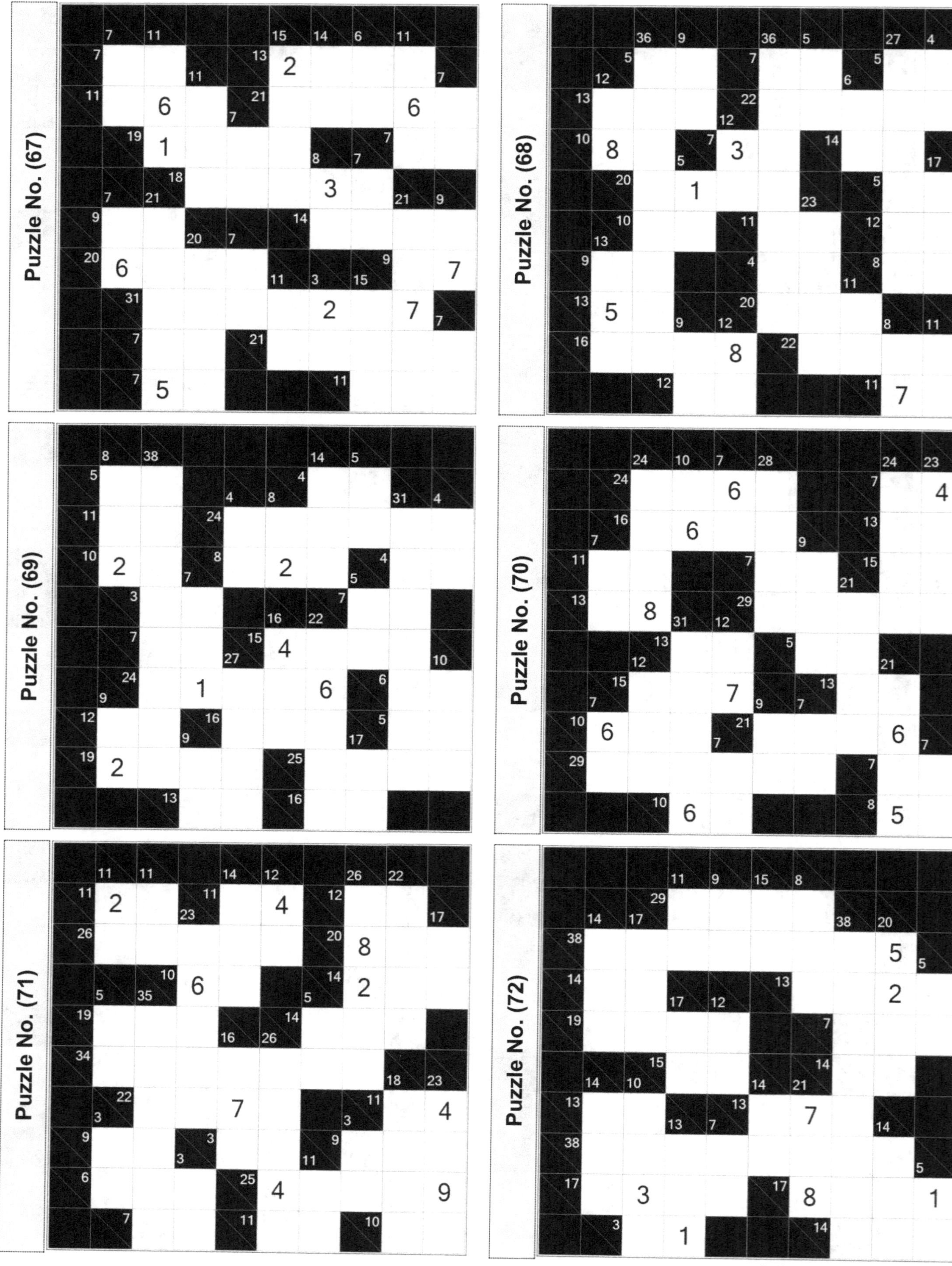

Solution on Page (173)

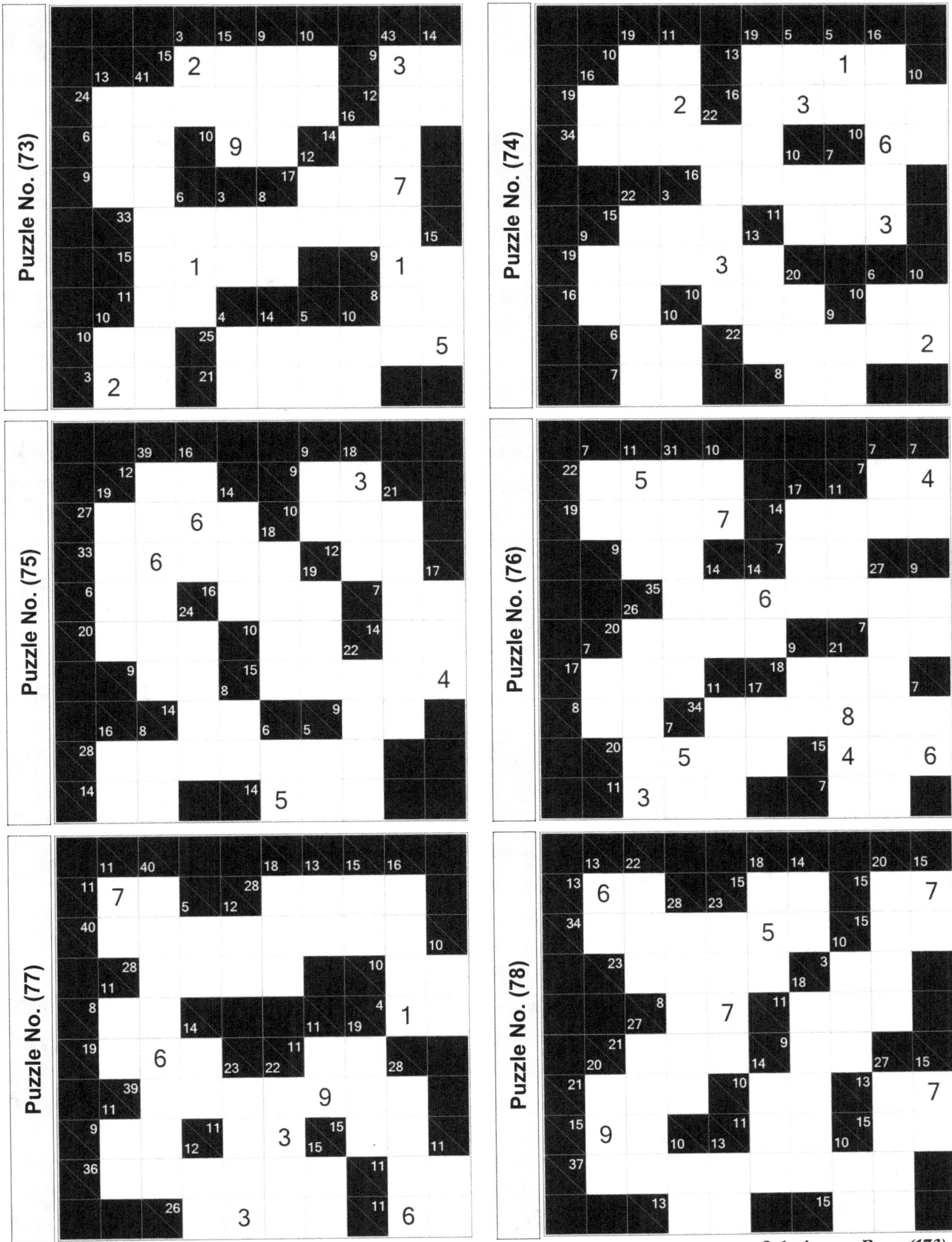

Solution on Page (173)

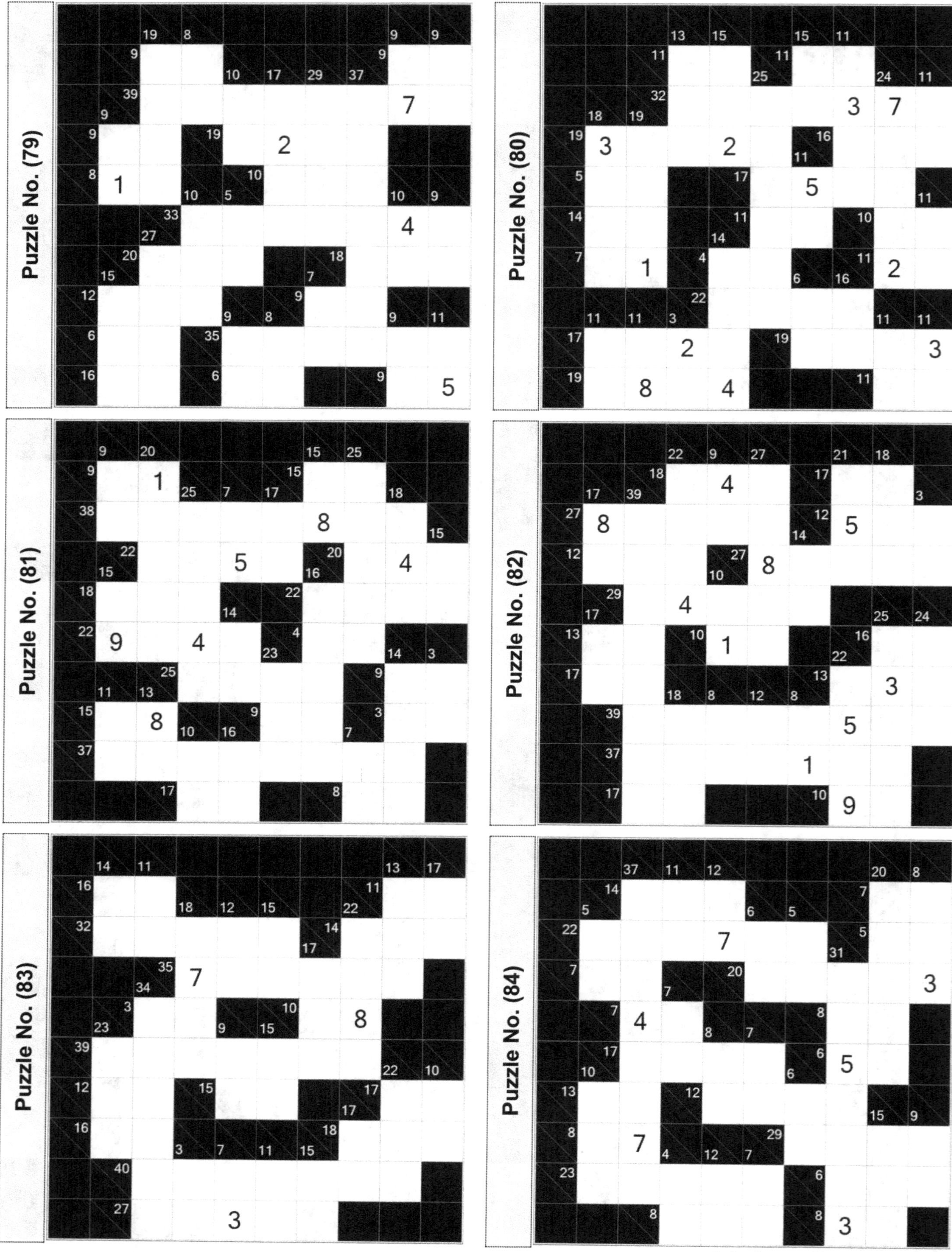

Solution on Pages (173-174)

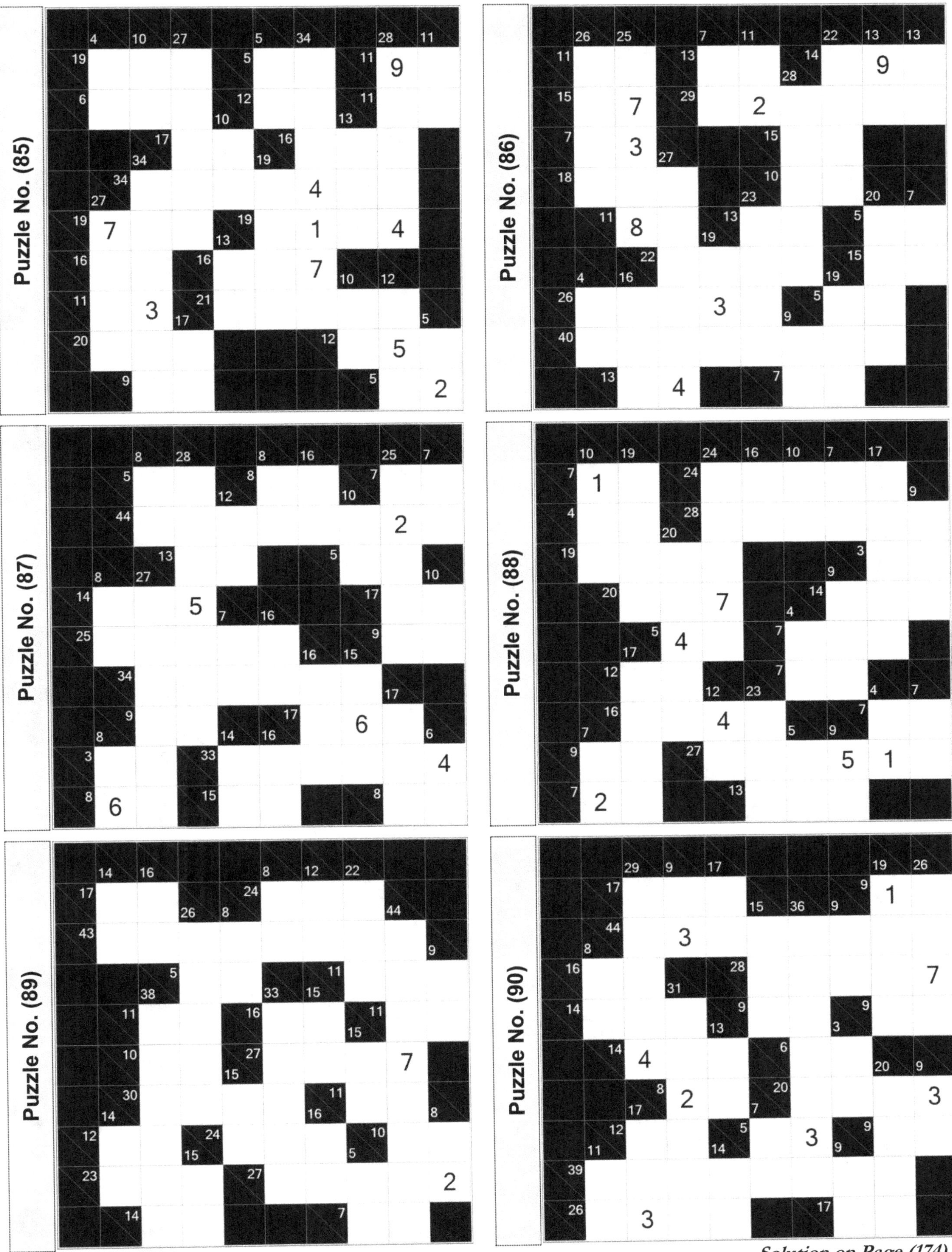

Solution on Page (174)

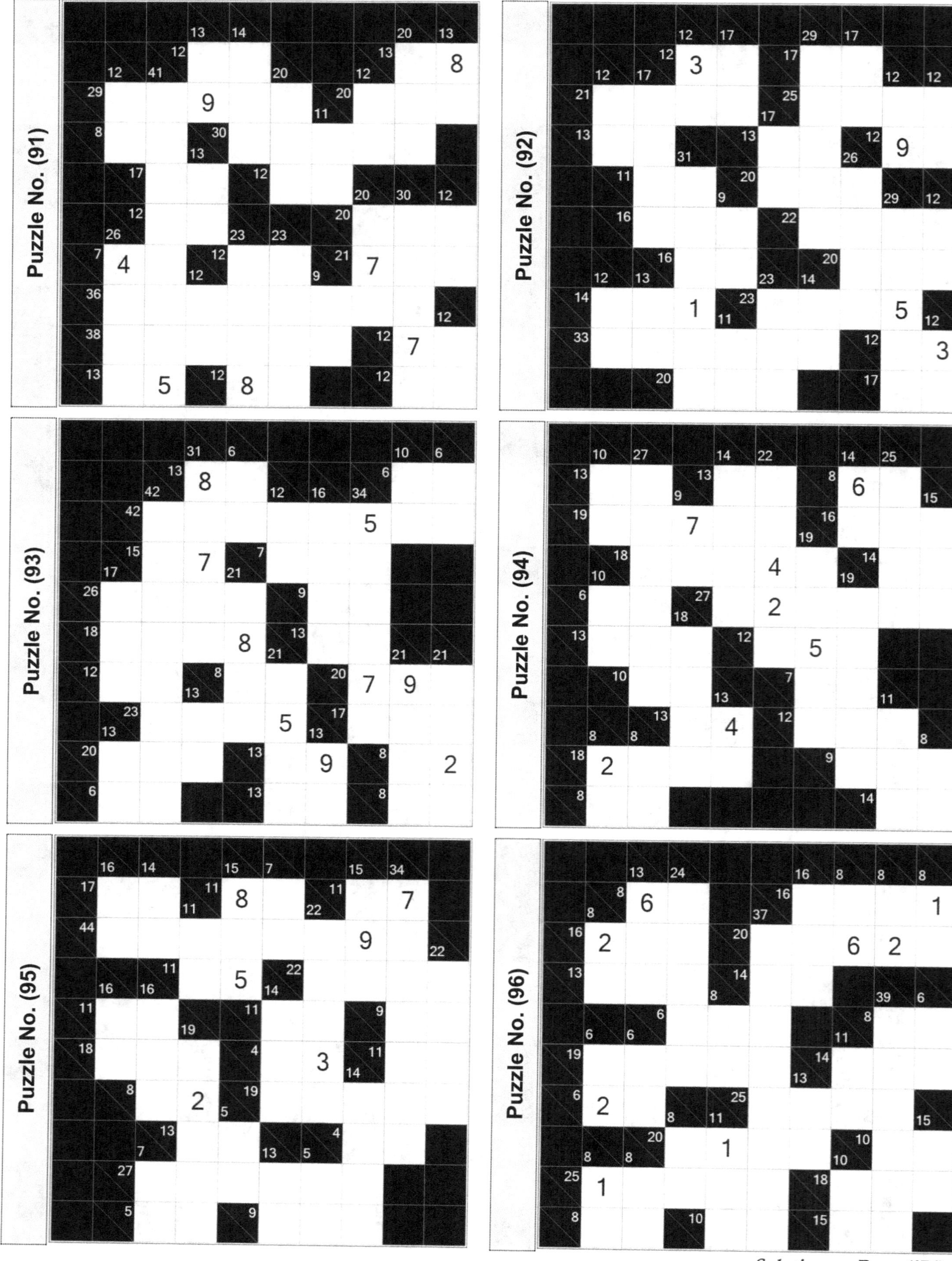

(18)

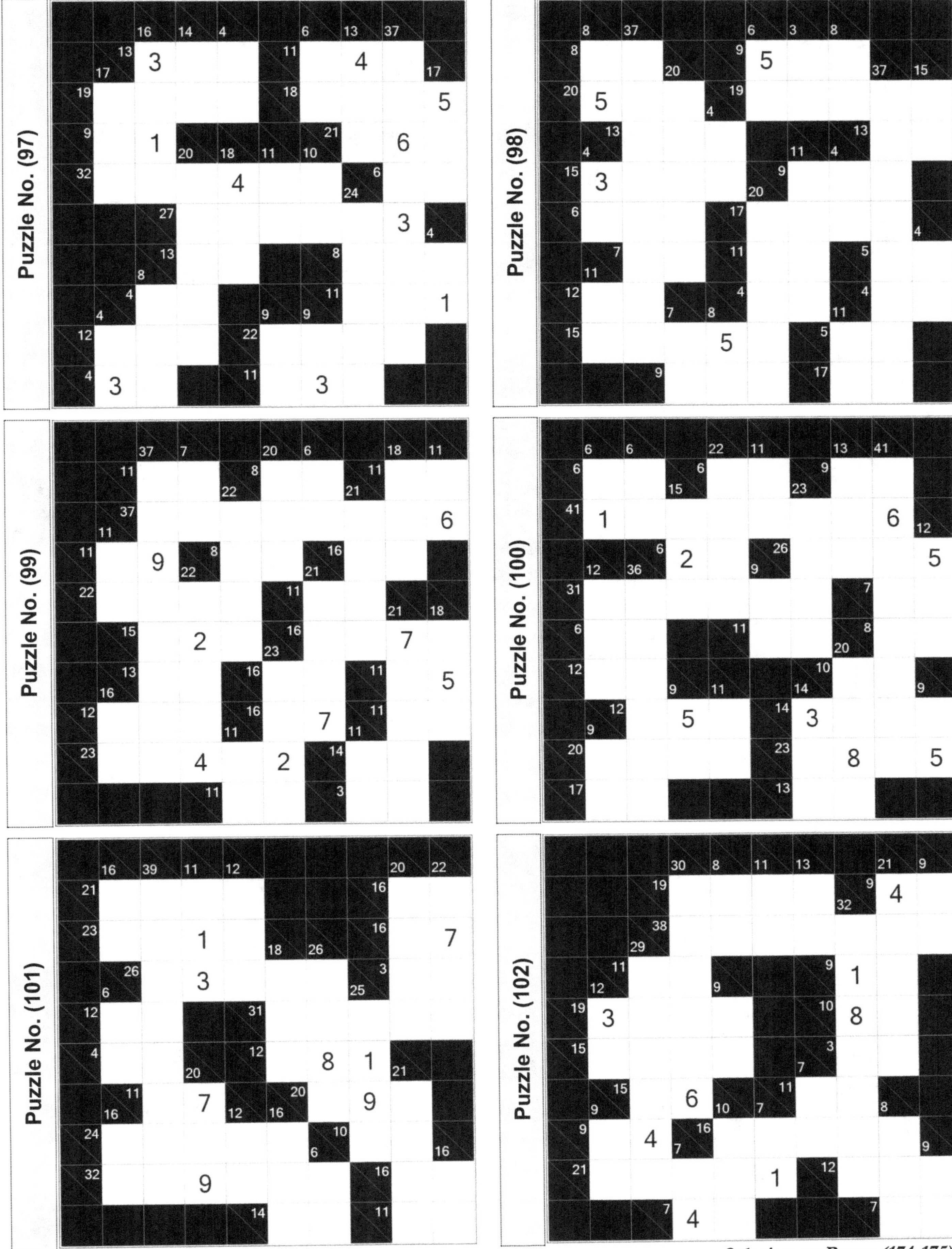

Solution on Pages (174-175)

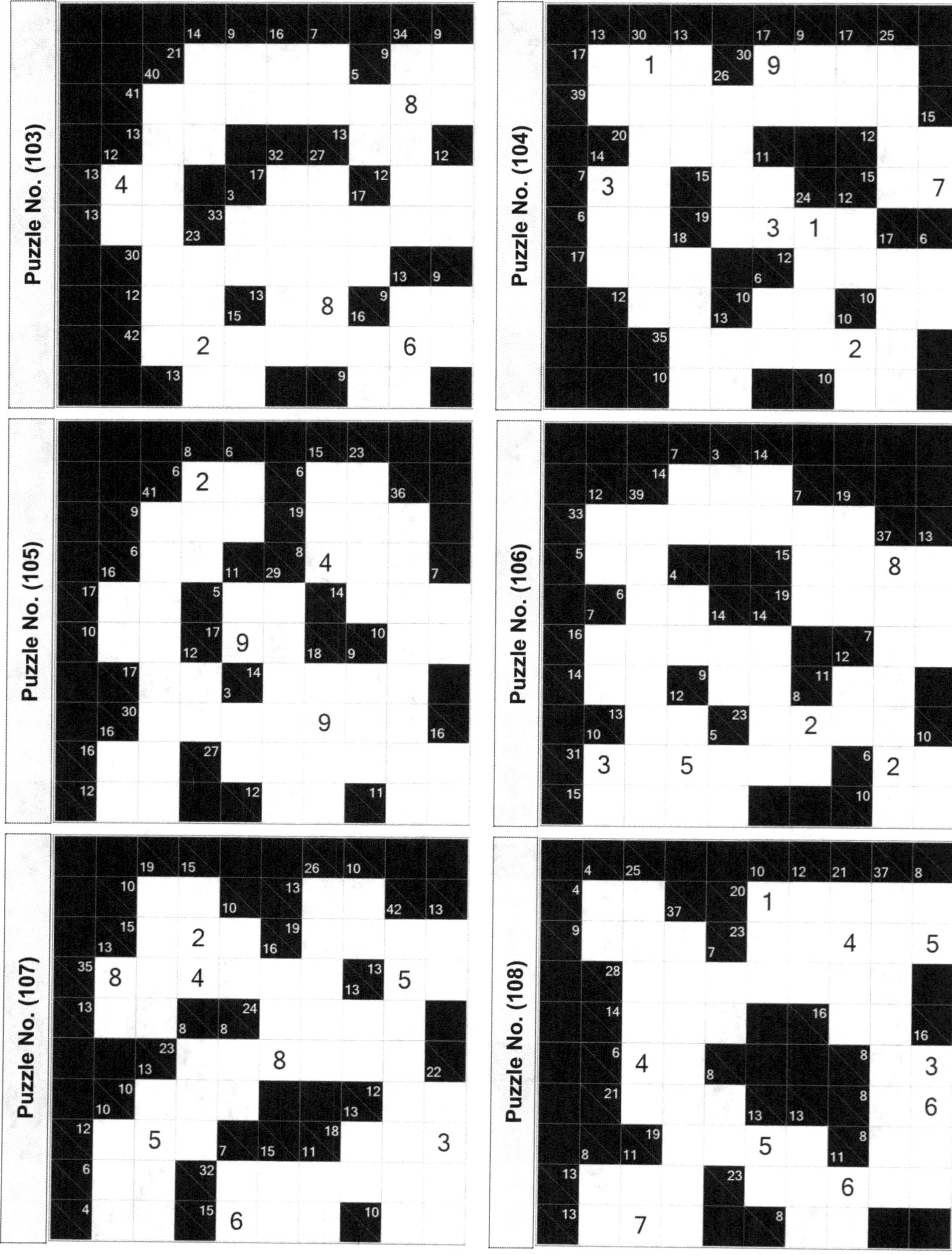

Solution on Page (175)

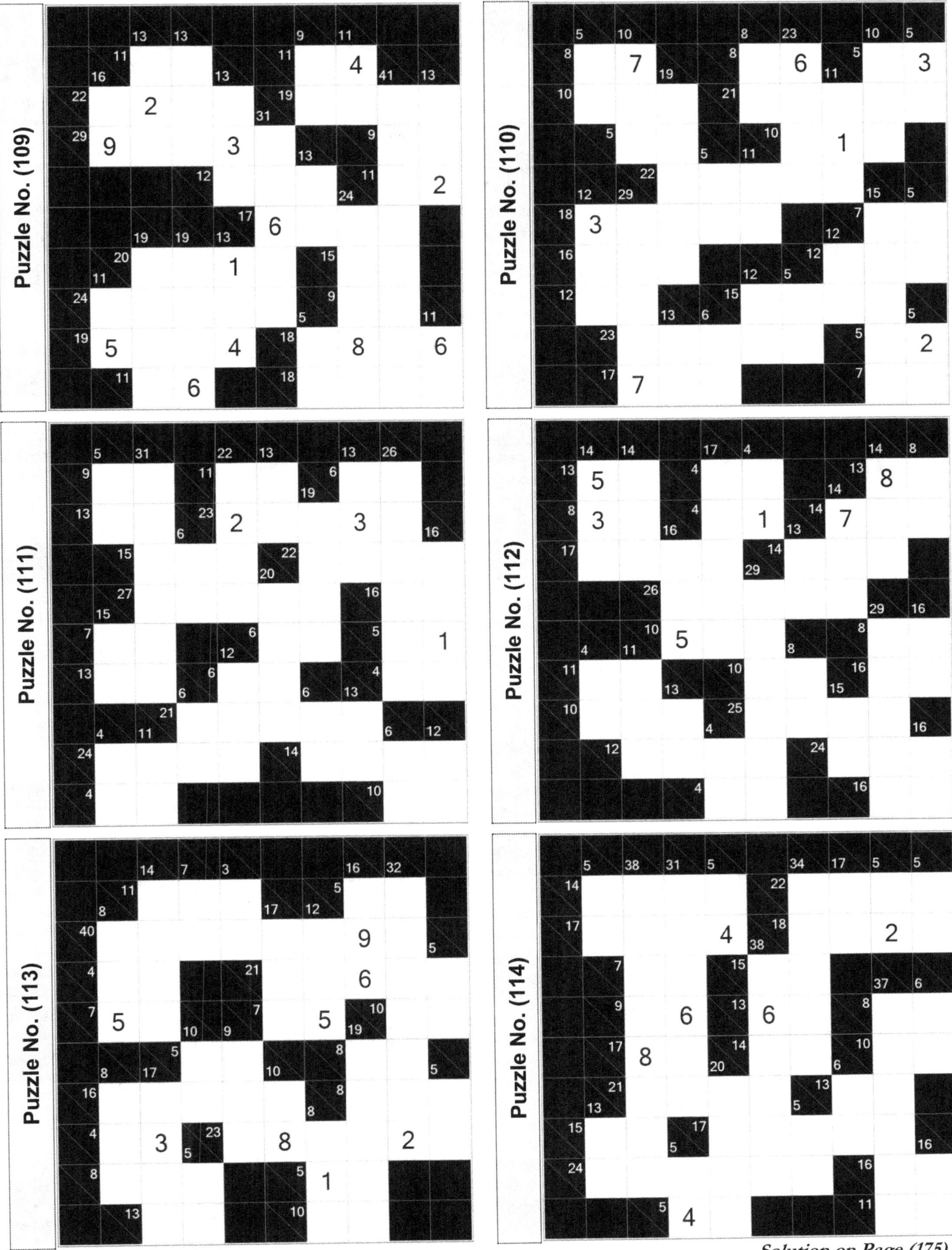

(21)

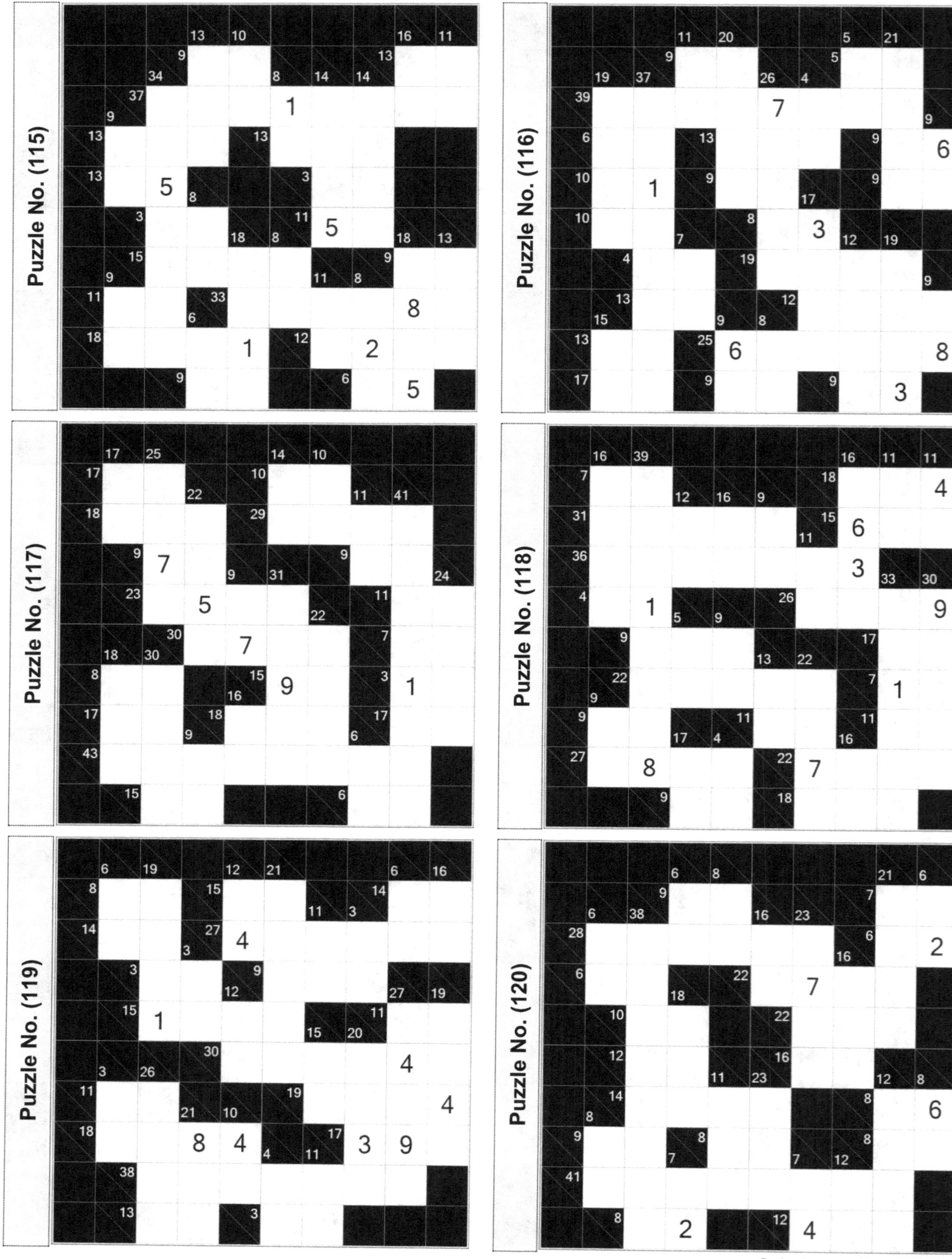

Solution on Page (175)

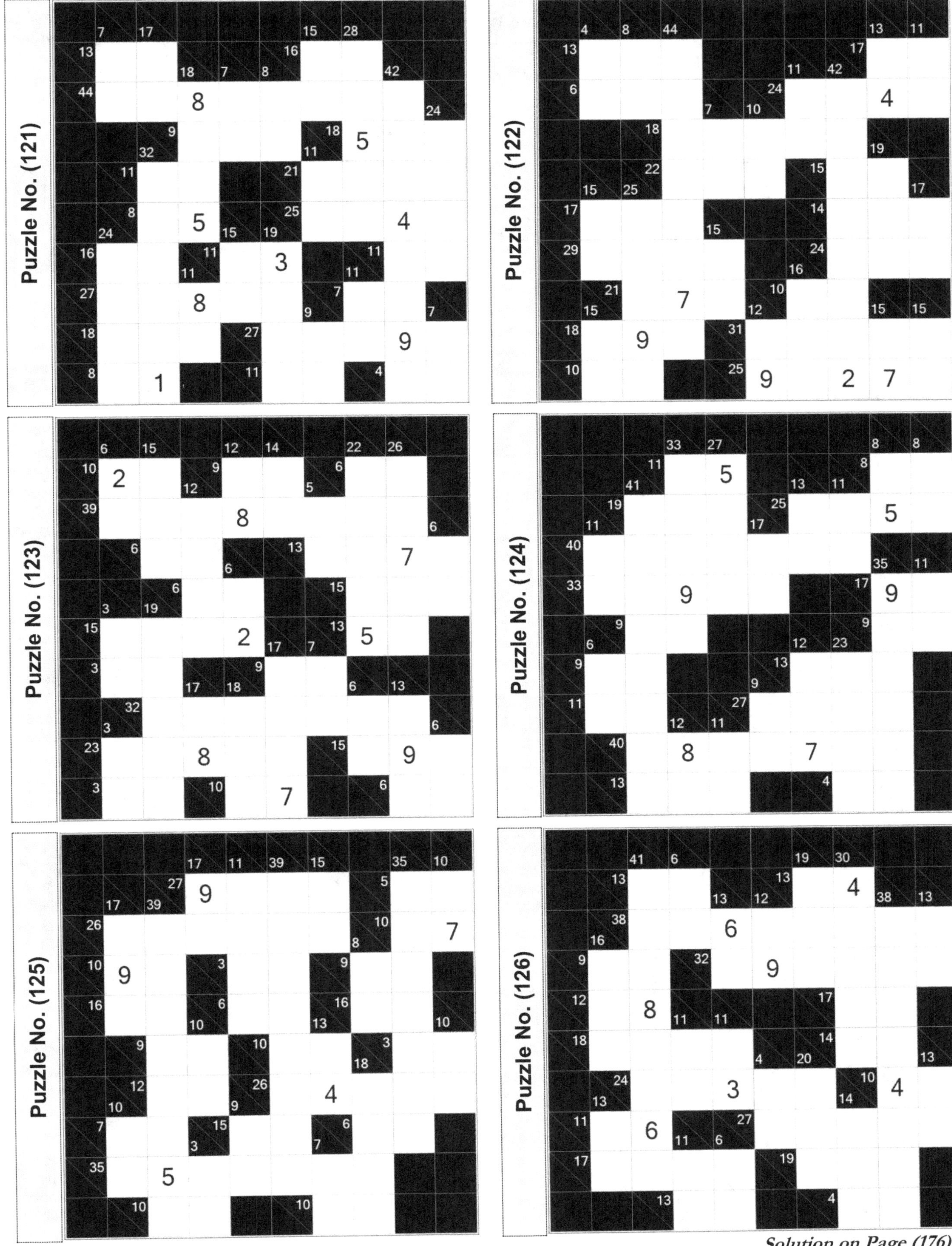

Solution on Page (176)

(23)

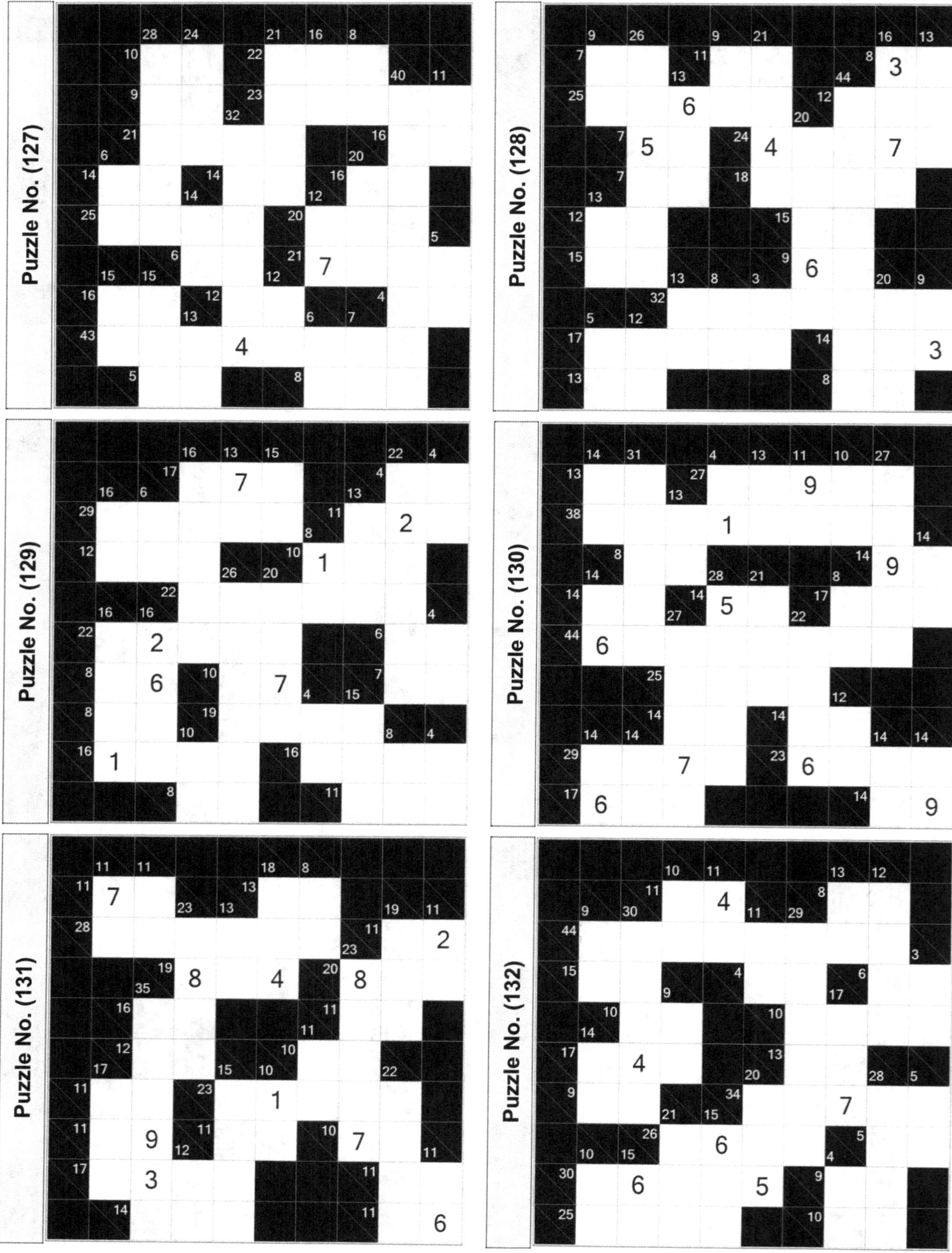

Solution on Page (176)

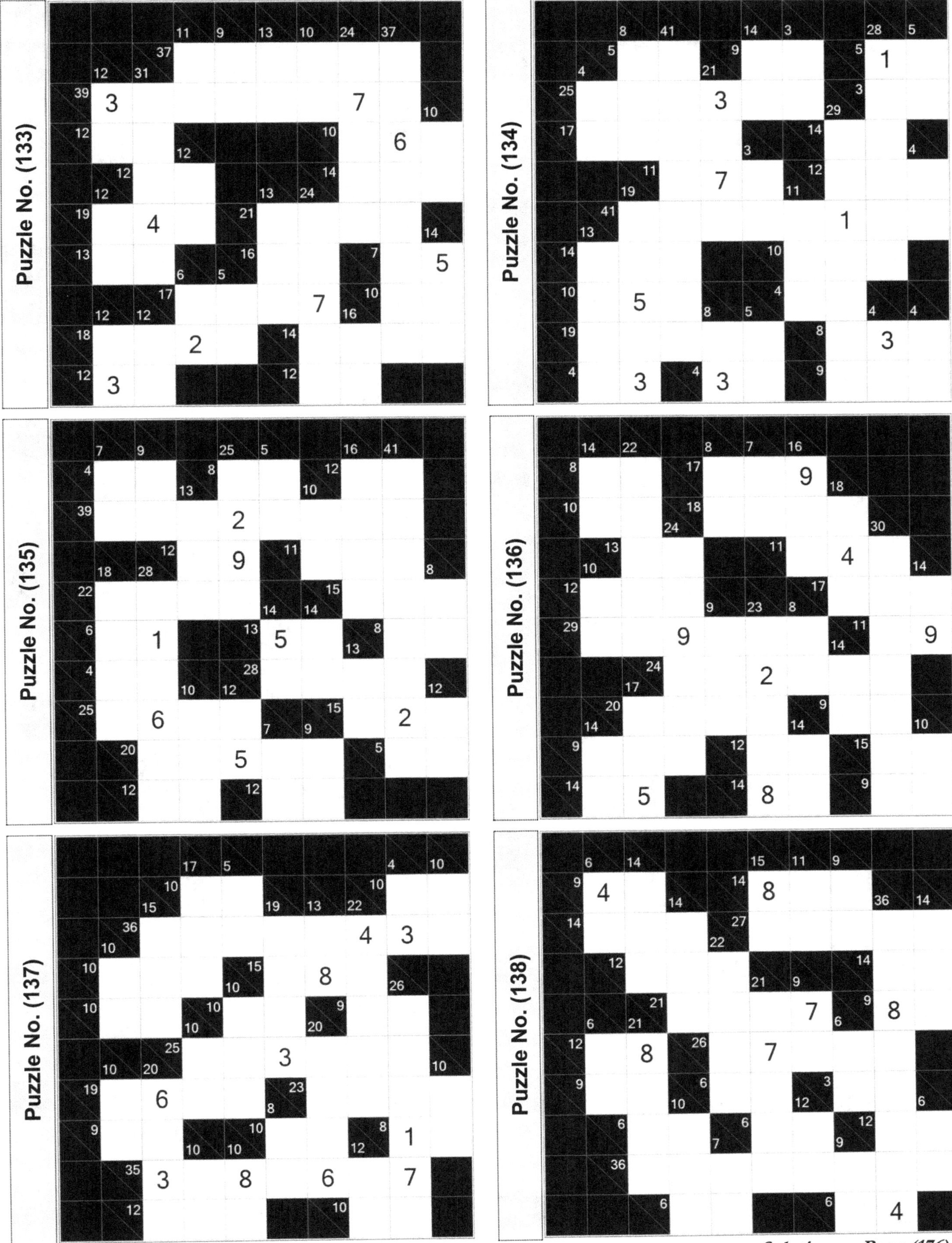

Solution on Page (176)

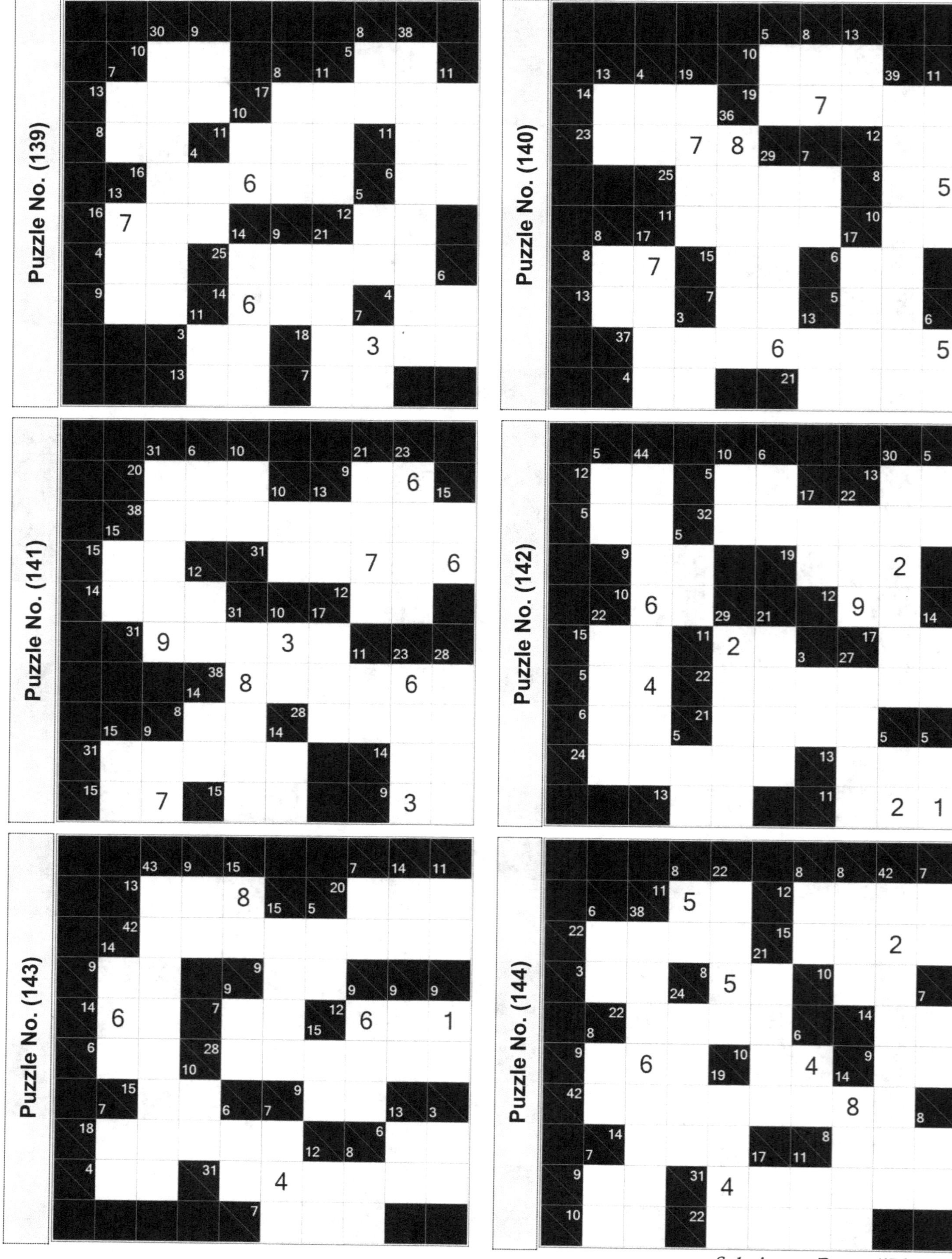

Puzzle No. (139)
Puzzle No. (140)
Puzzle No. (141)
Puzzle No. (142)
Puzzle No. (143)
Puzzle No. (144)

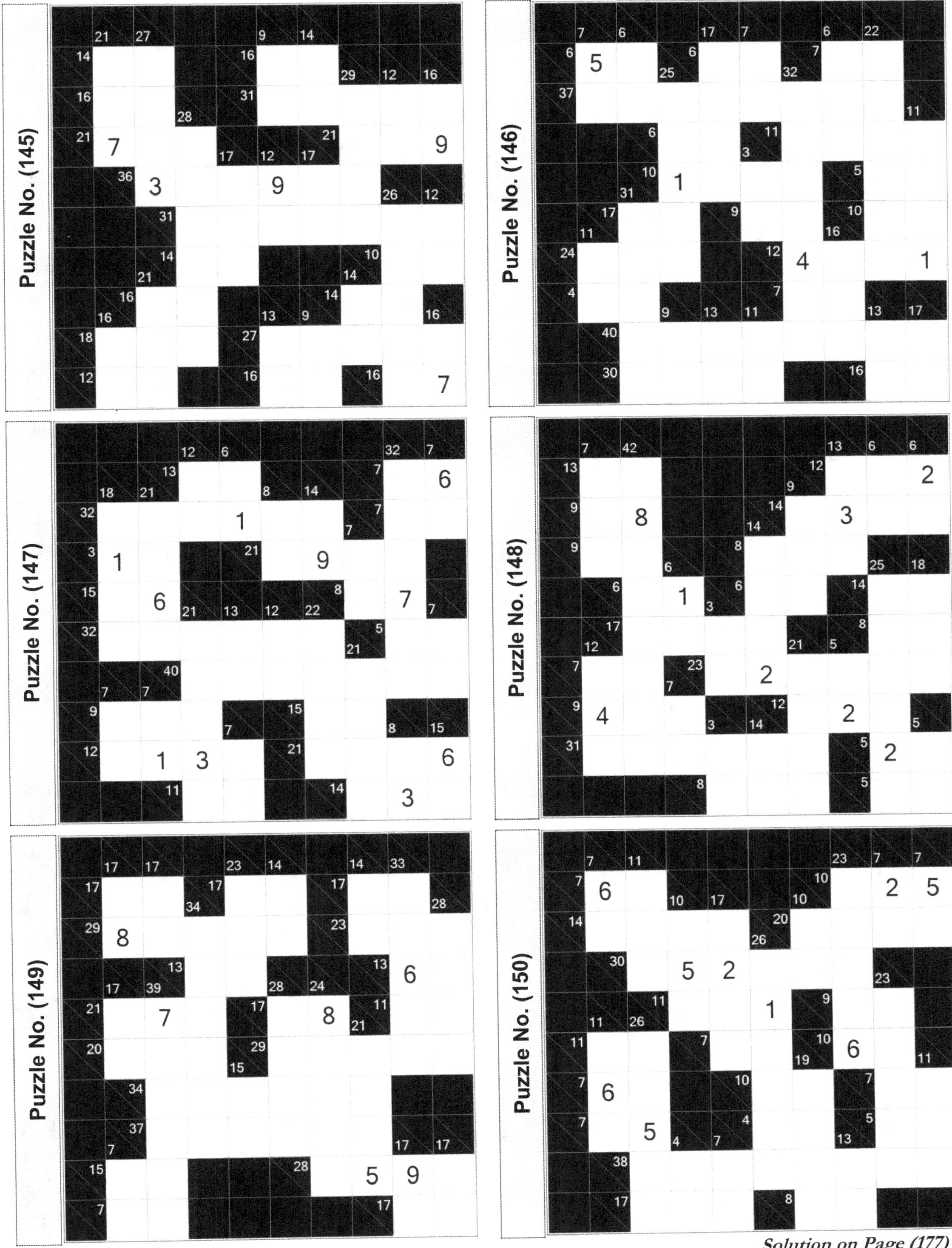

Solution on Page (177)

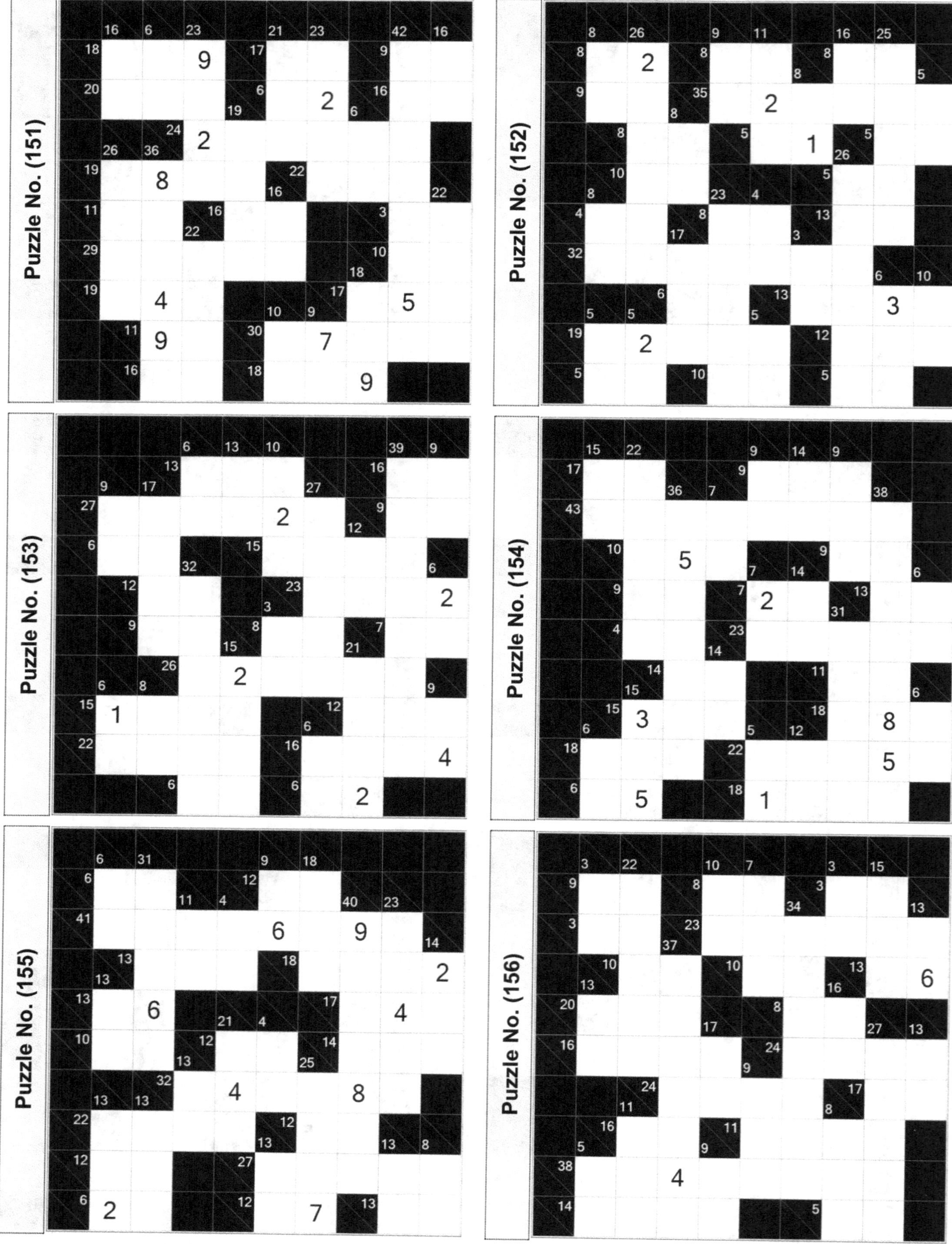

Puzzle No. (151)

Puzzle No. (152)

Puzzle No. (153)

Puzzle No. (154)

Puzzle No. (155)

Puzzle No. (156)

Solution on Page (177)

Puzzle No. (157)
Puzzle No. (158)
Puzzle No. (159)
Puzzle No. (160)
Puzzle No. (161)
Puzzle No. (162)
Solution on Pages (177-178)

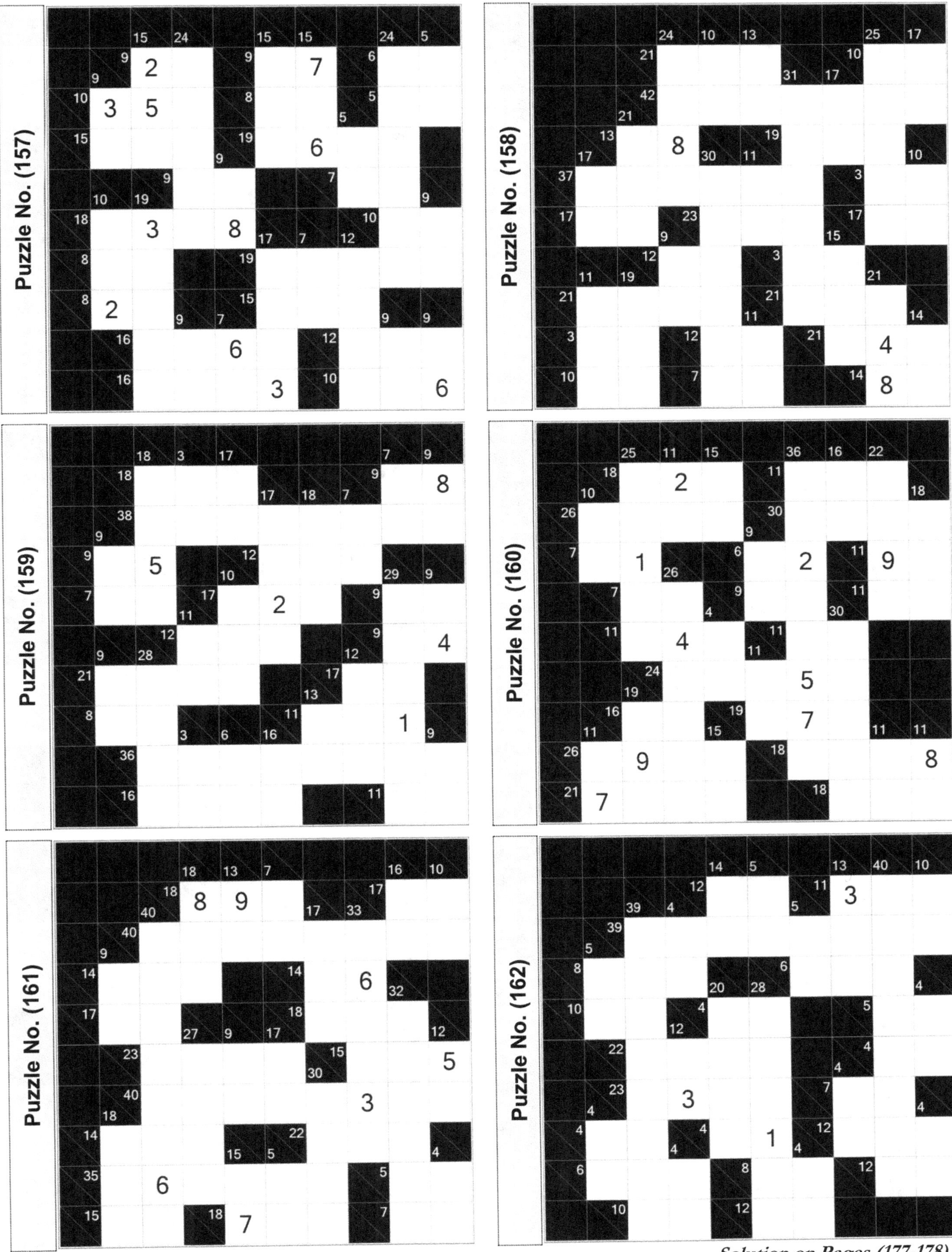

(29)

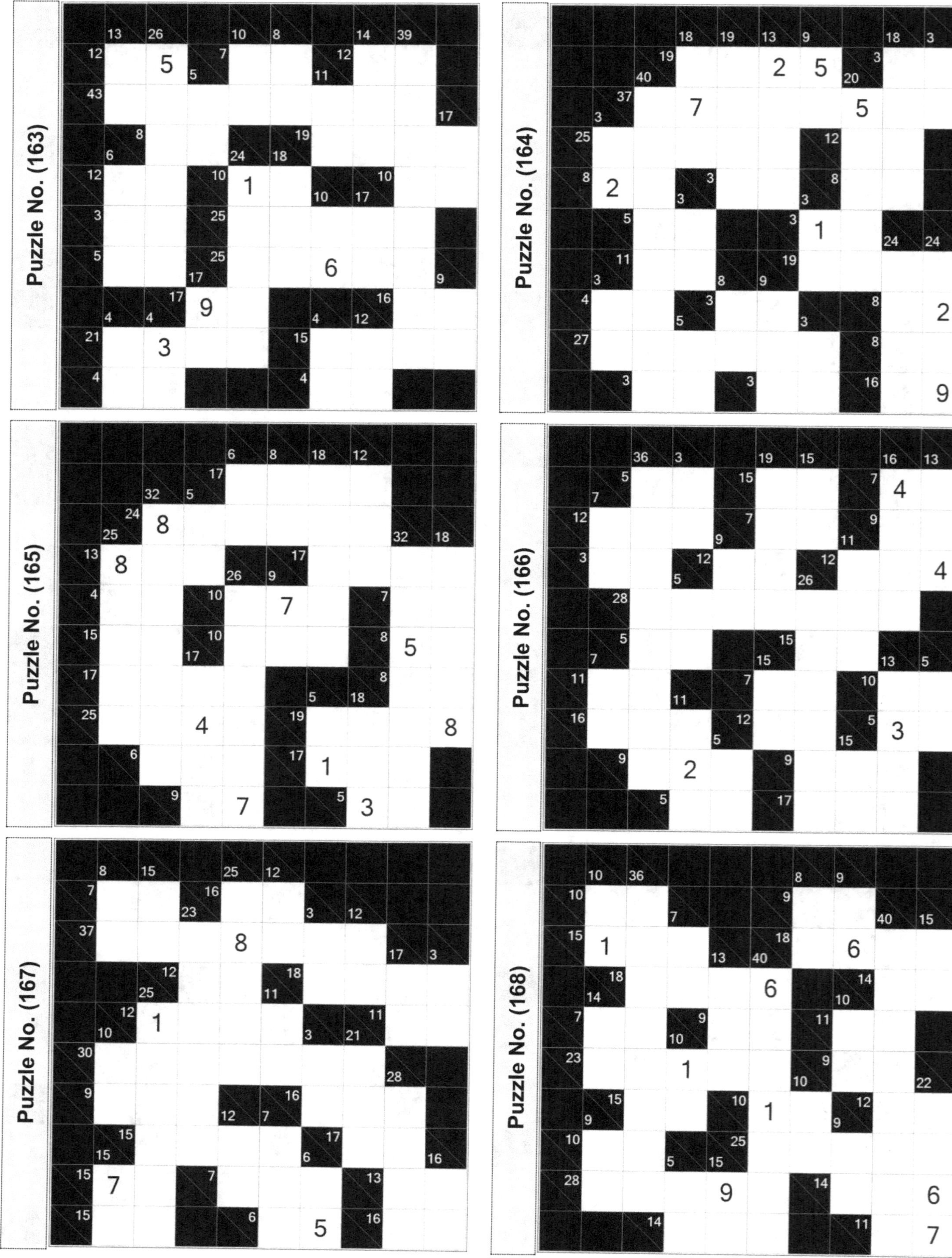

Puzzle No. (163)

Puzzle No. (164)

Puzzle No. (165)

Puzzle No. (166)

Puzzle No. (167)

Puzzle No. (168)

Solution on Page (178)

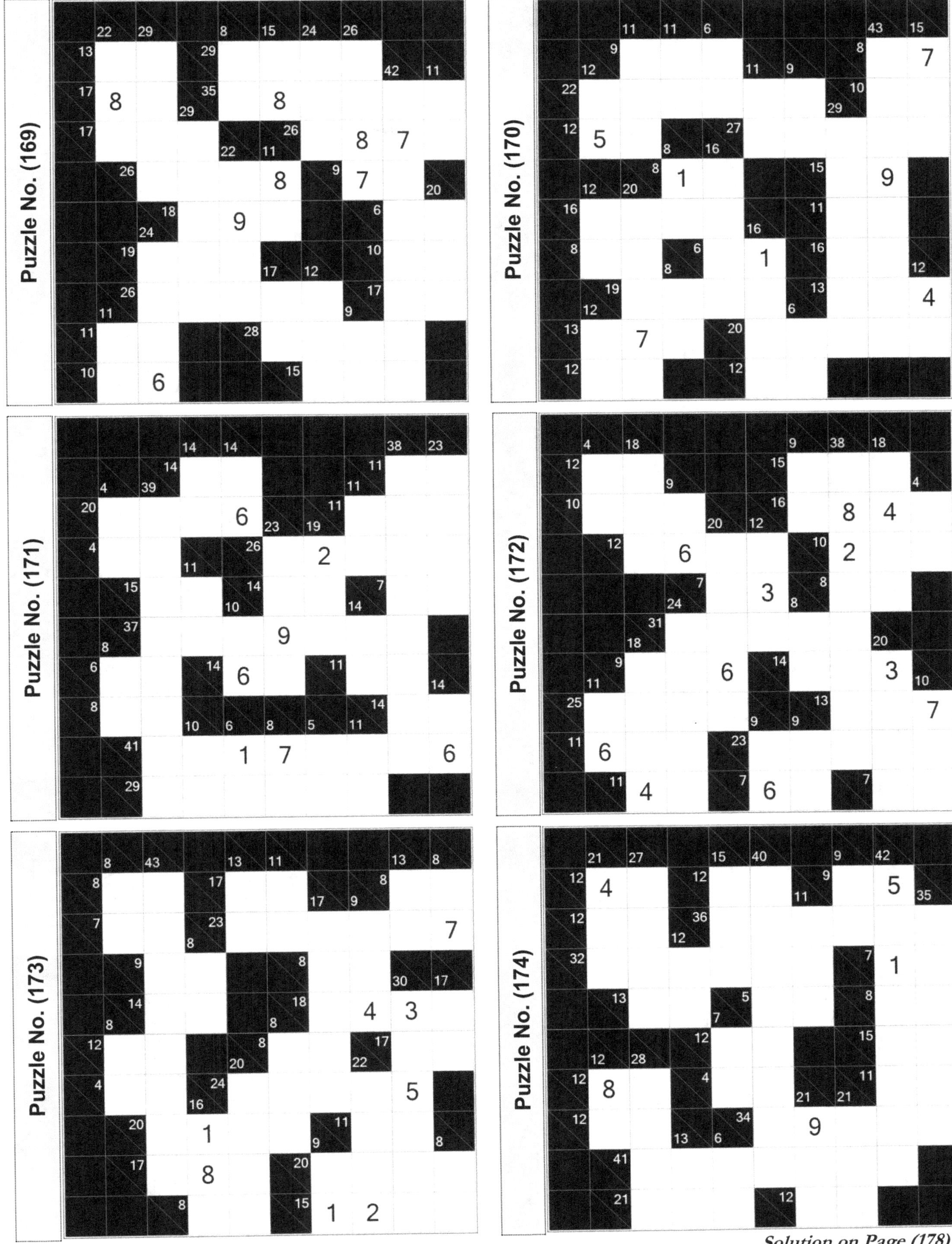

Puzzle No. (169)
Puzzle No. (170)
Puzzle No. (171)
Puzzle No. (172)
Puzzle No. (173)
Puzzle No. (174)
Solution on Page (178)

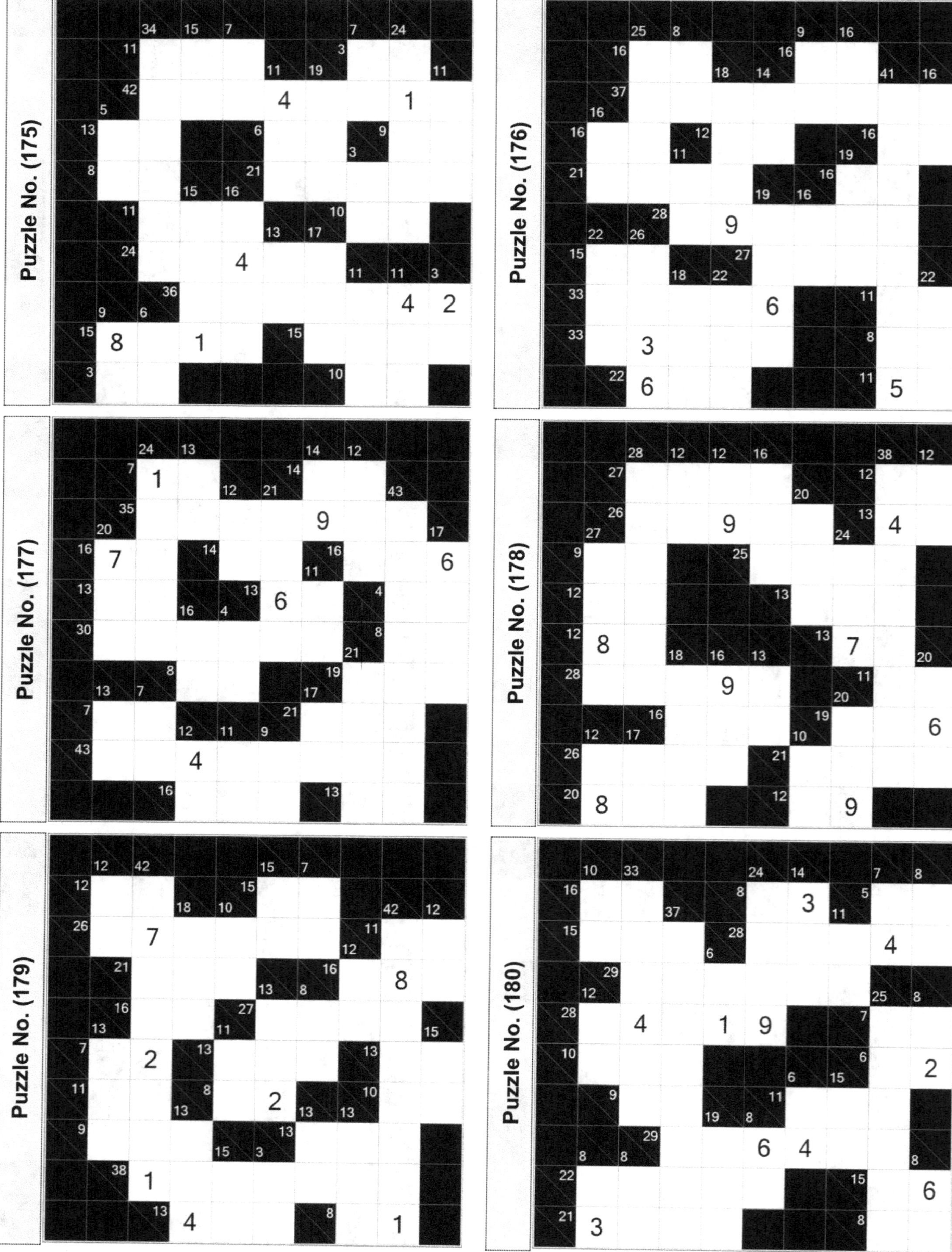

Solution on Page (178)

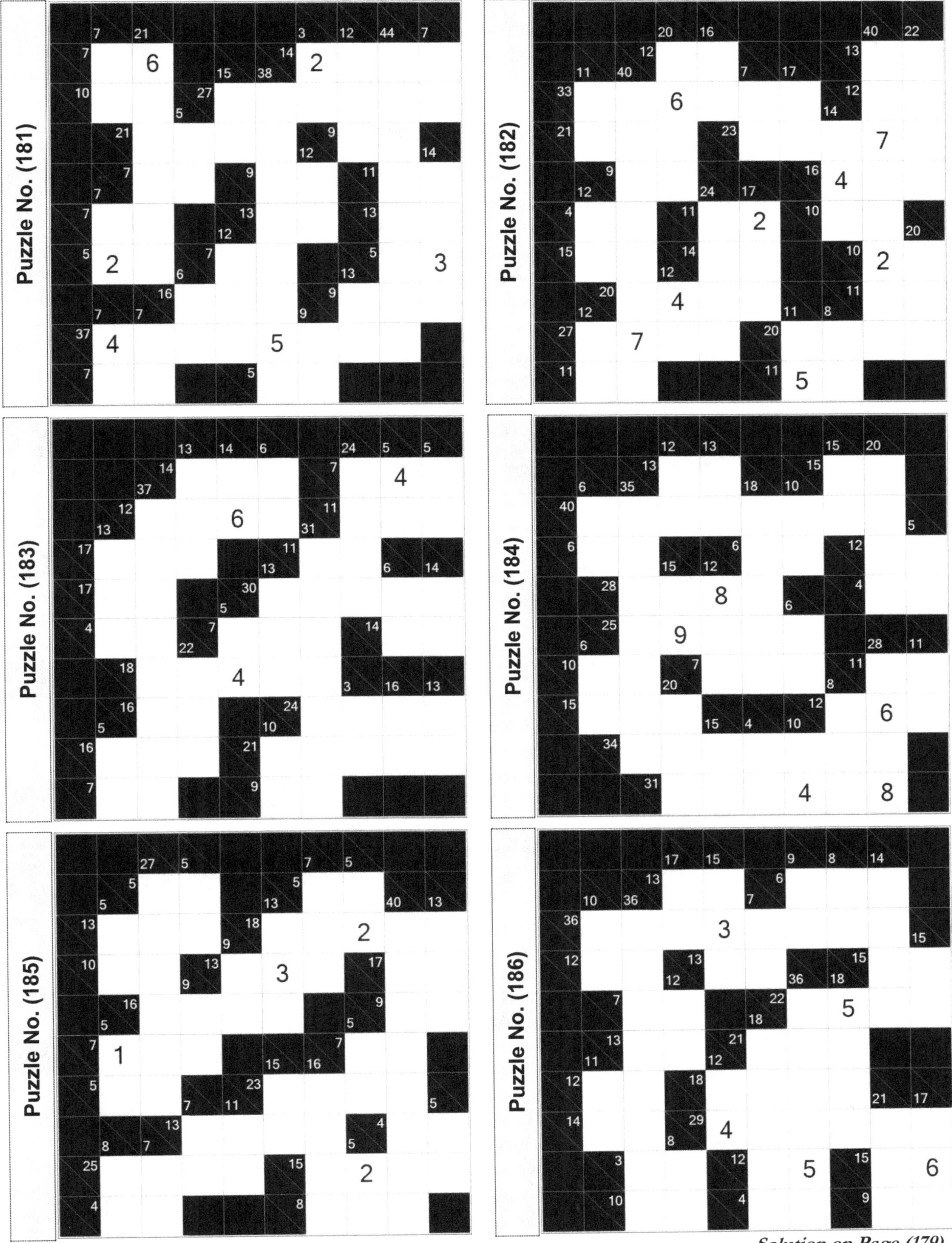

Solution on Page (179)

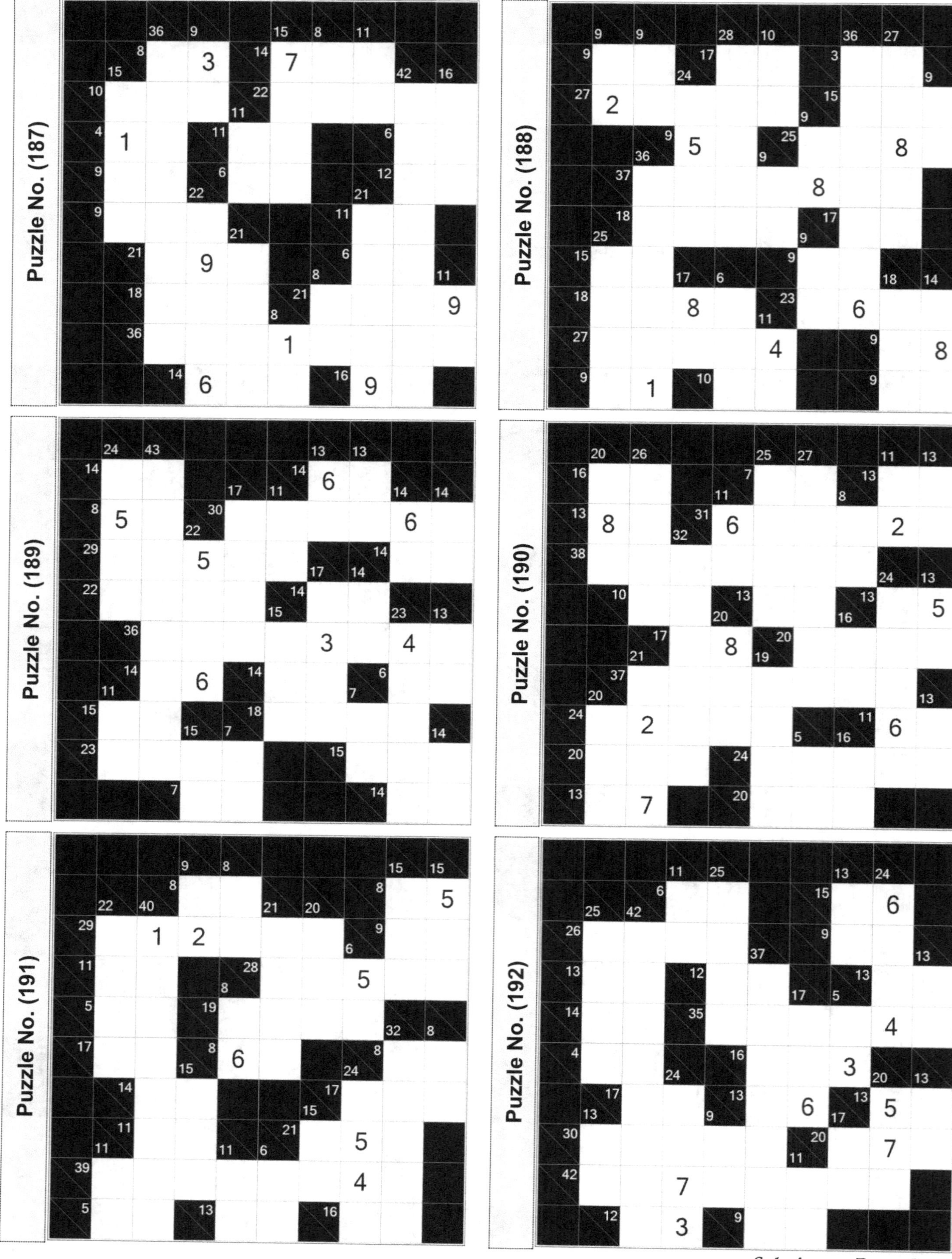

Puzzle No. (187)

Puzzle No. (188)

Puzzle No. (189)

Puzzle No. (190)

Puzzle No. (191)

Puzzle No. (192)

Solution on Page (179)

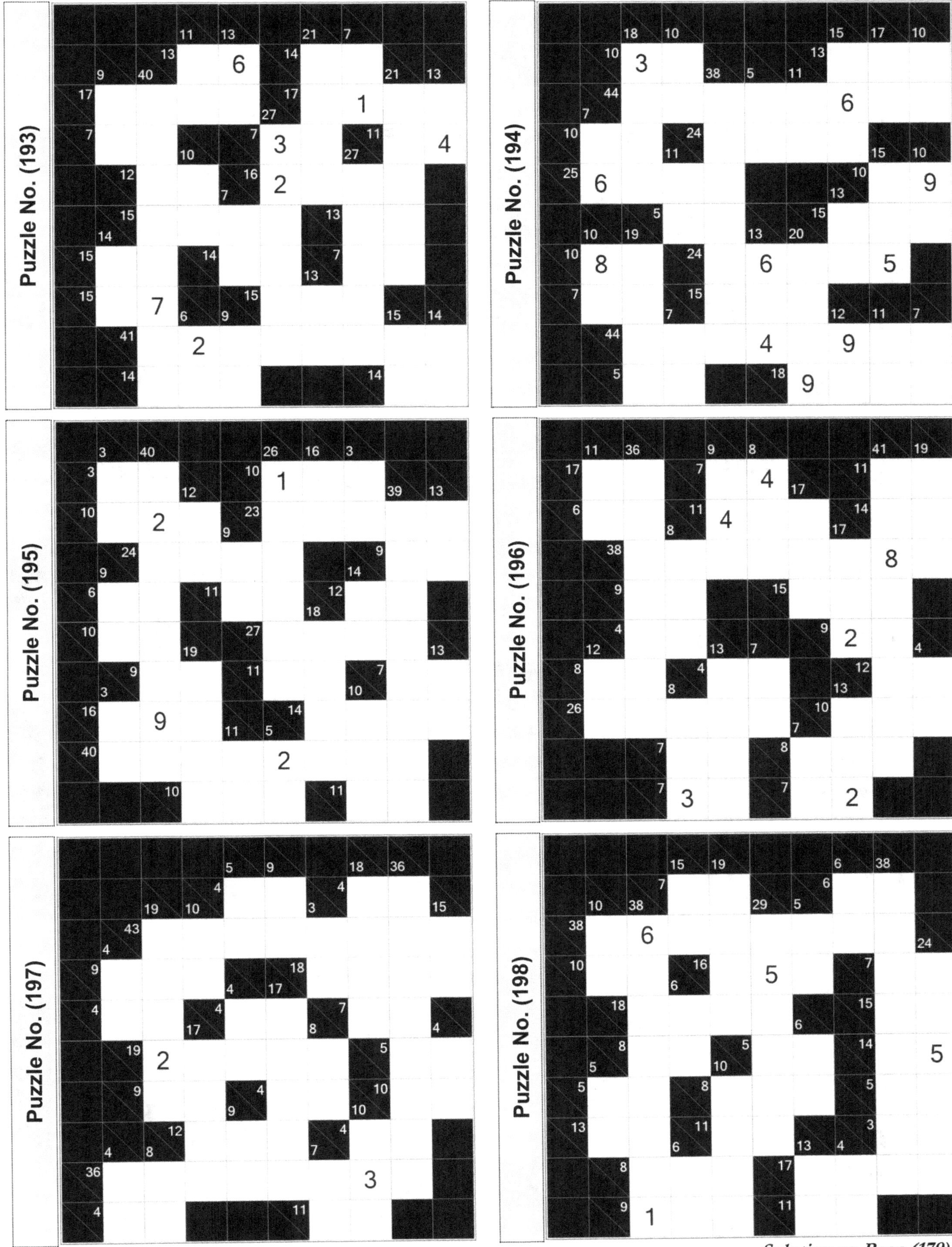

Solution on Page (179)

(35)

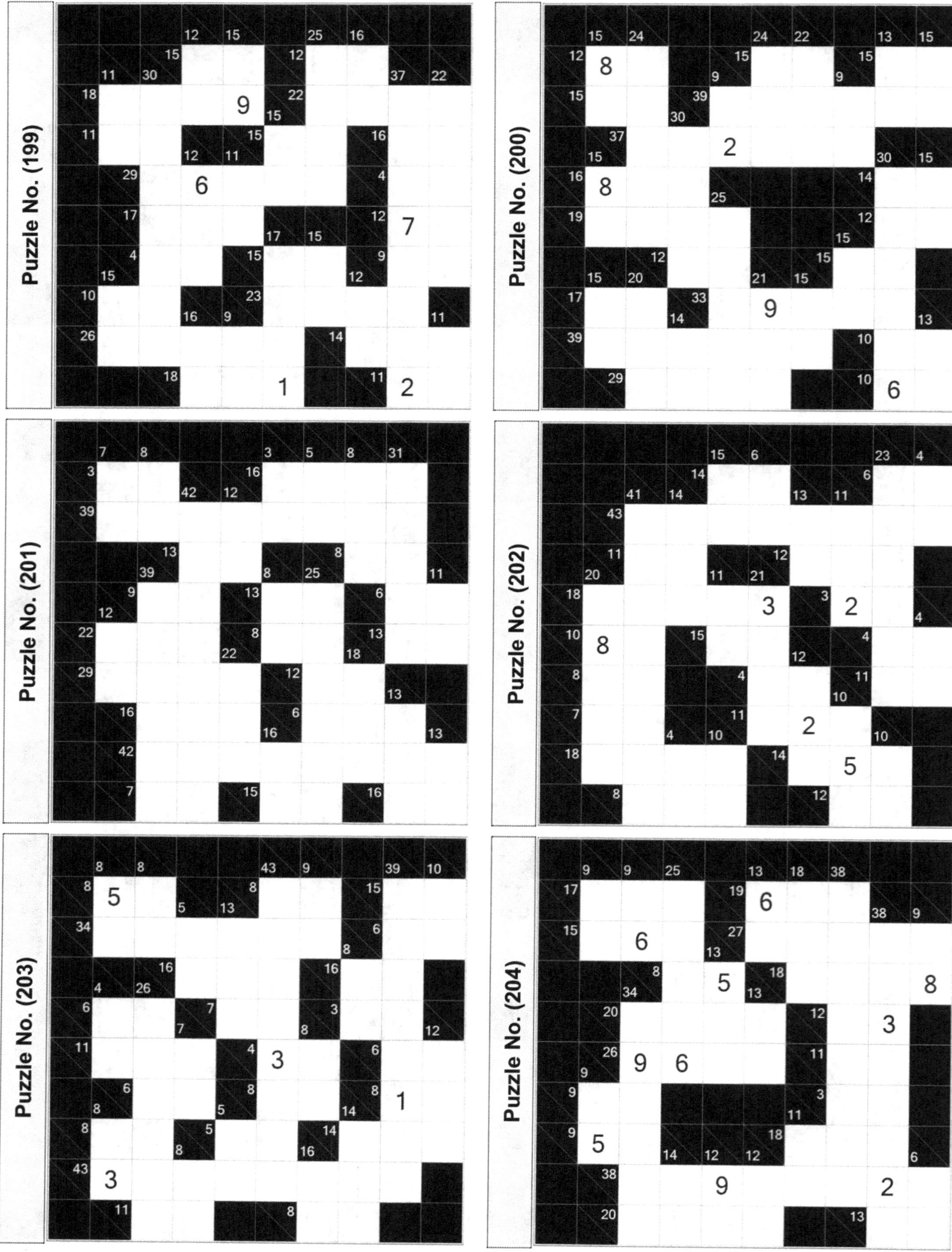

Puzzle No. (199)

Puzzle No. (200)

Puzzle No. (201)

Puzzle No. (202)

Puzzle No. (203)

Puzzle No. (204)

Solution on Pages (179-180)

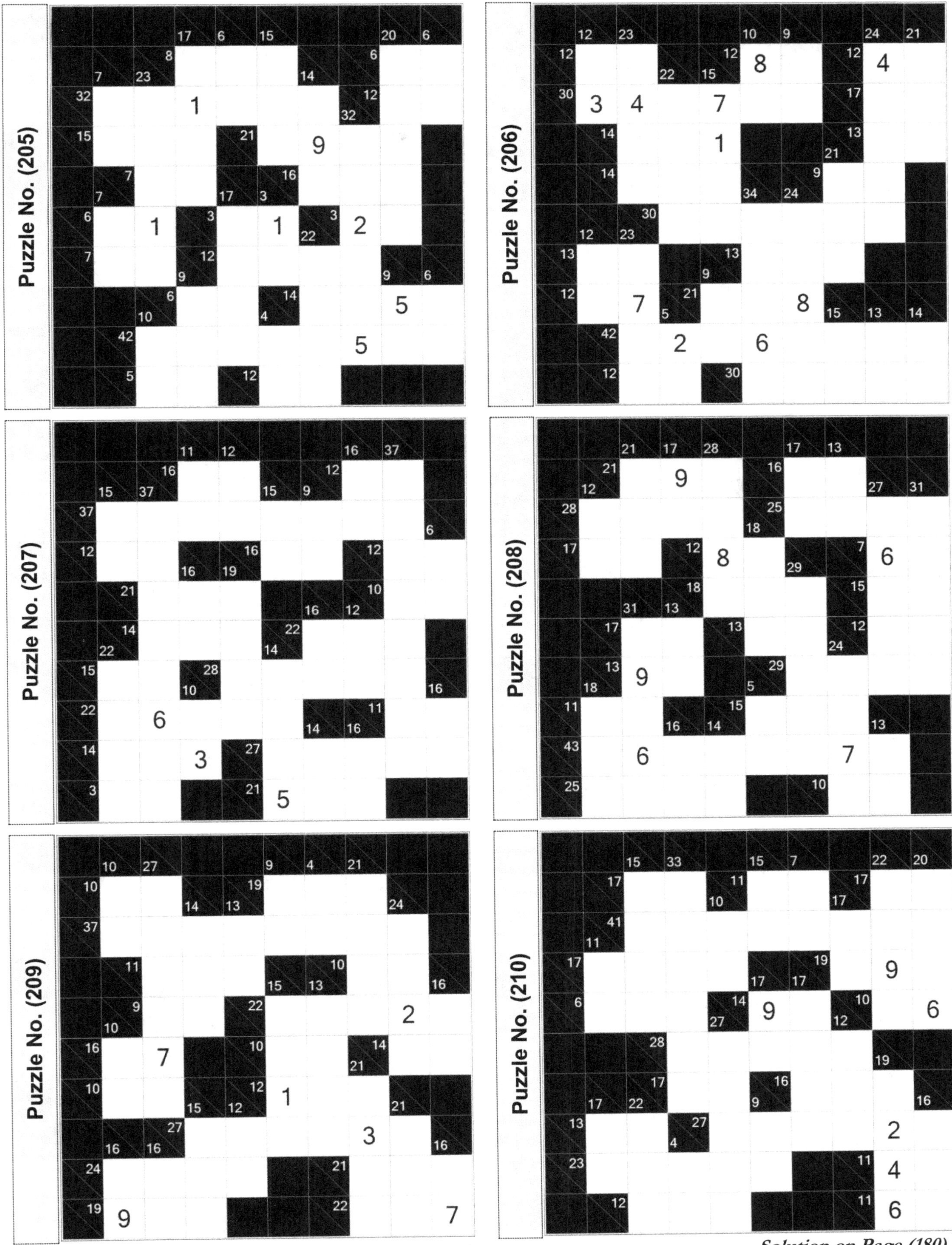

Solution on Page (180)

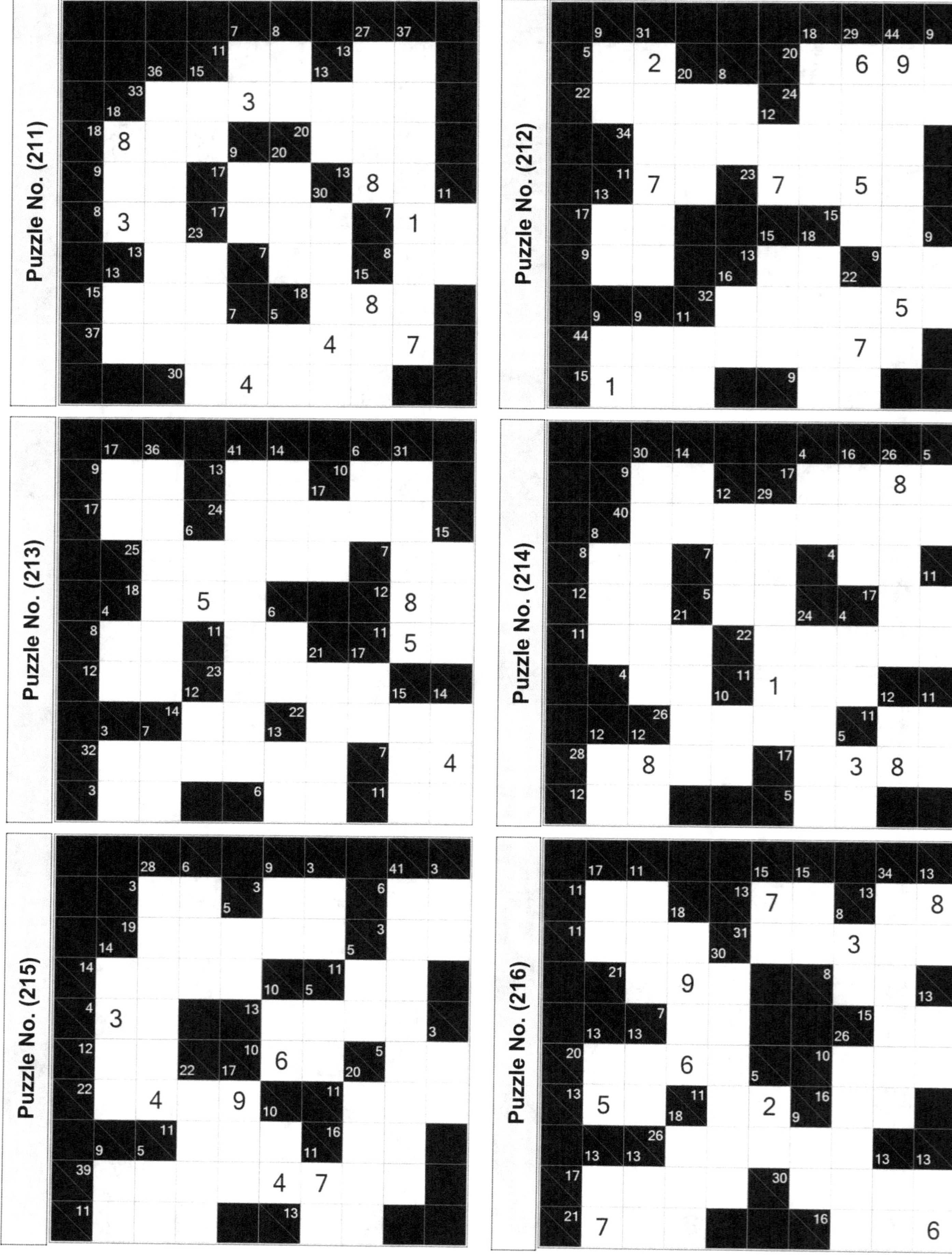

Puzzle No. (211)
Puzzle No. (212)
Puzzle No. (213)
Puzzle No. (214)
Puzzle No. (215)
Puzzle No. (216)

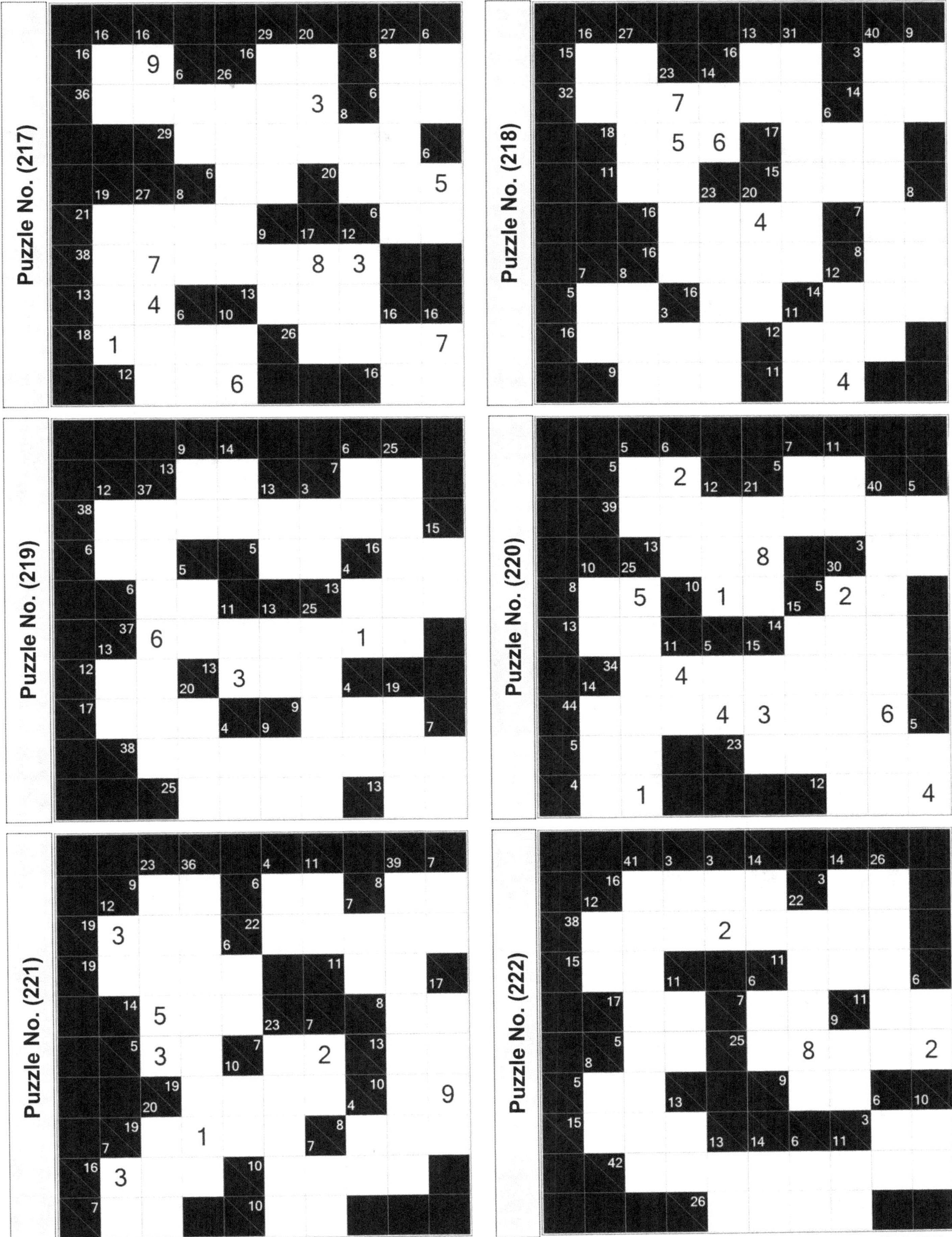

Solution on Pages (180-181)

(39)

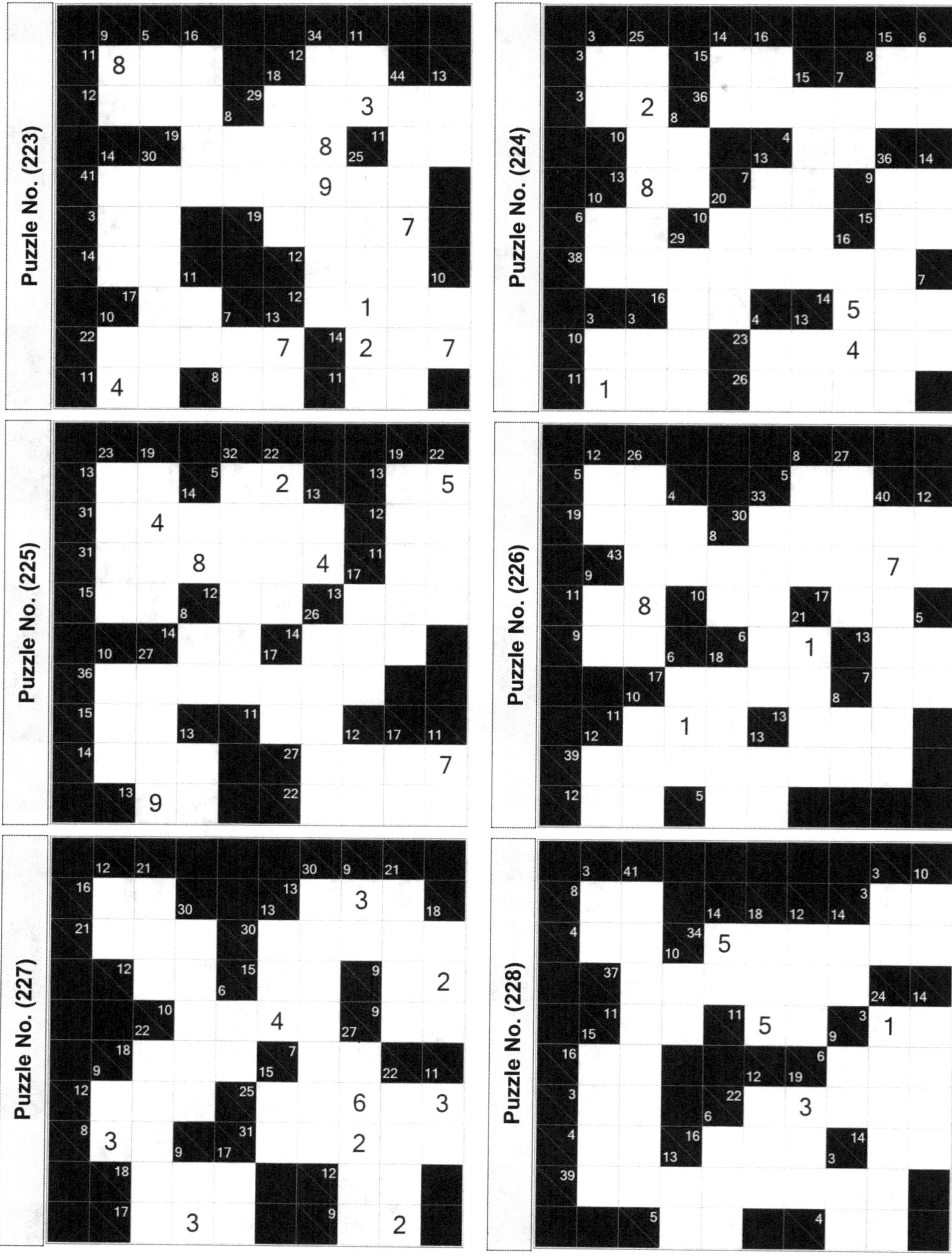

Solution on Page (181)

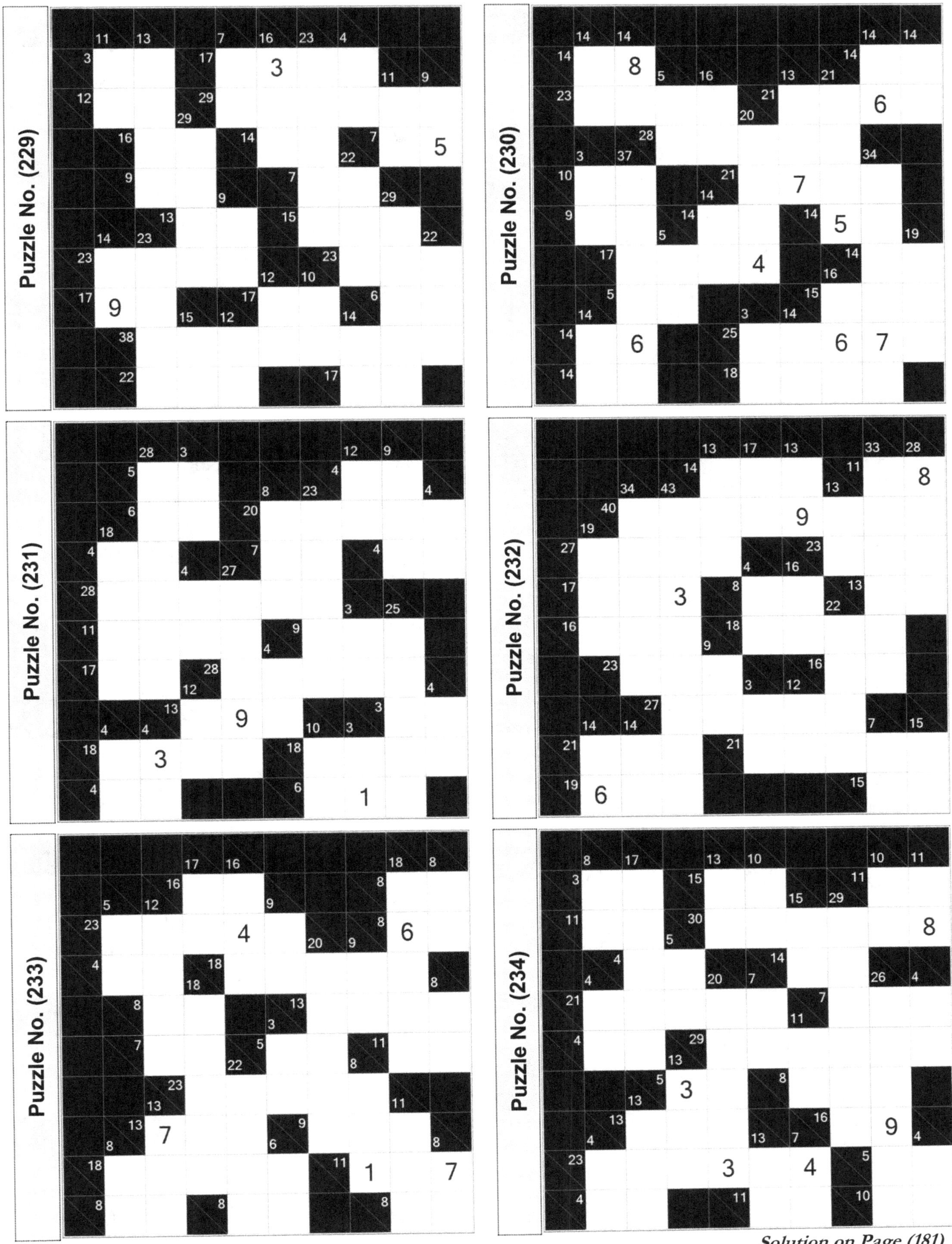

Puzzle No. (229)

Puzzle No. (230)

Puzzle No. (231)

Puzzle No. (232)

Puzzle No. (233)

Puzzle No. (234)

Solution on Page (181)

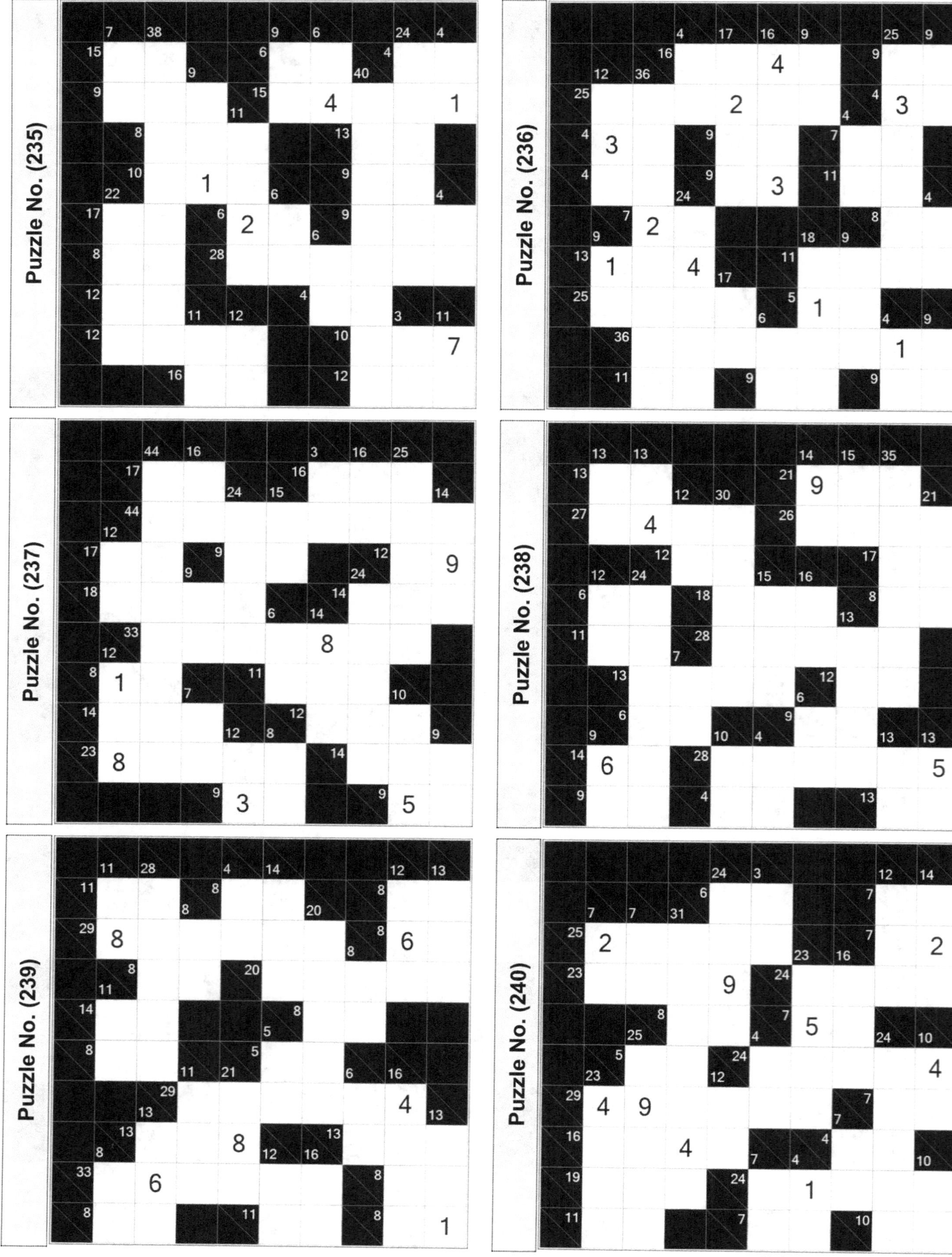

Puzzle No. (235)

Puzzle No. (236)

Puzzle No. (237)

Puzzle No. (238)

Puzzle No. (239)

Puzzle No. (240)

Solution on Page (181)

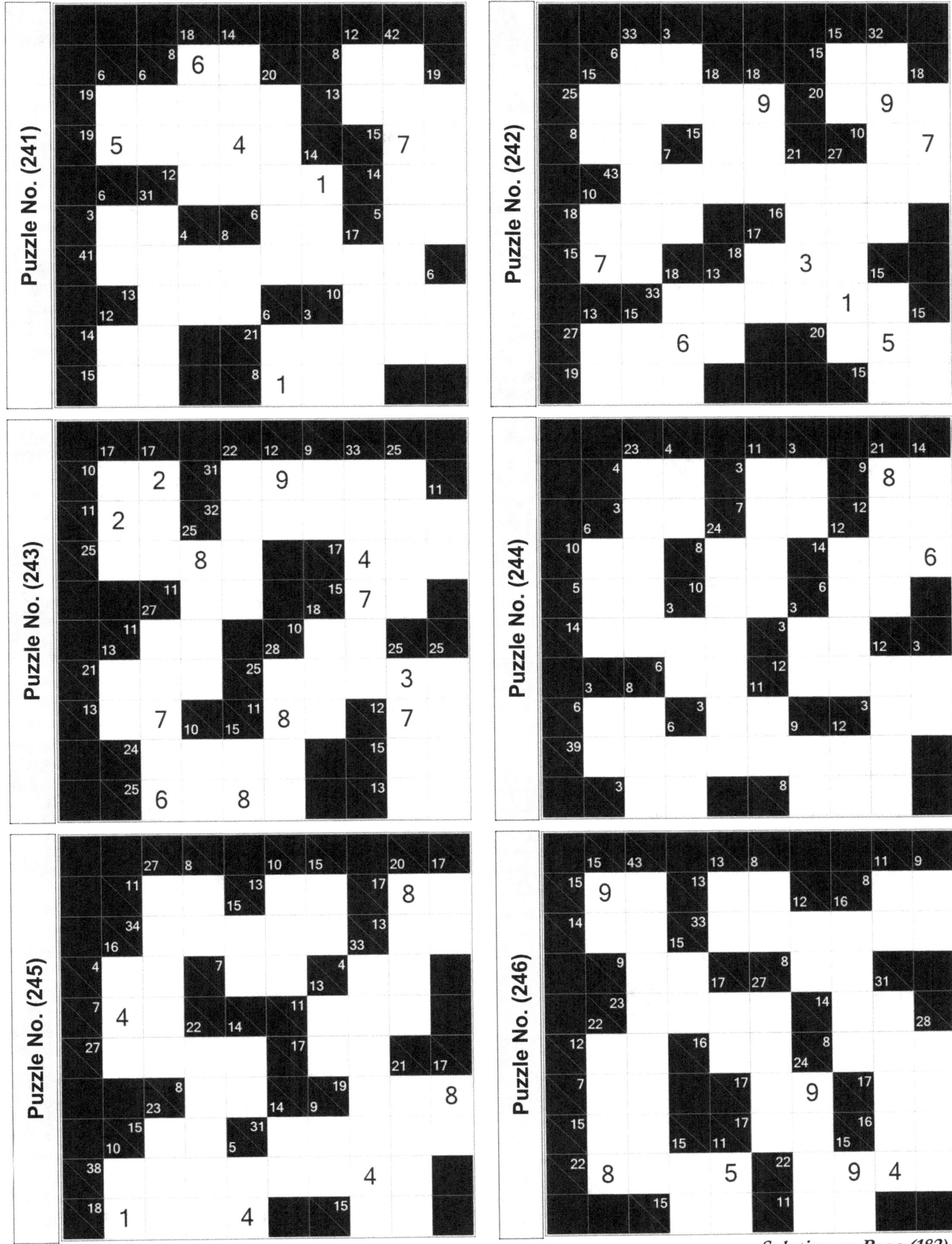

Puzzle No. (241)

Puzzle No. (242)

Puzzle No. (243)

Puzzle No. (244)

Puzzle No. (245)

Puzzle No. (246)

Solution on Page (182)

(43)

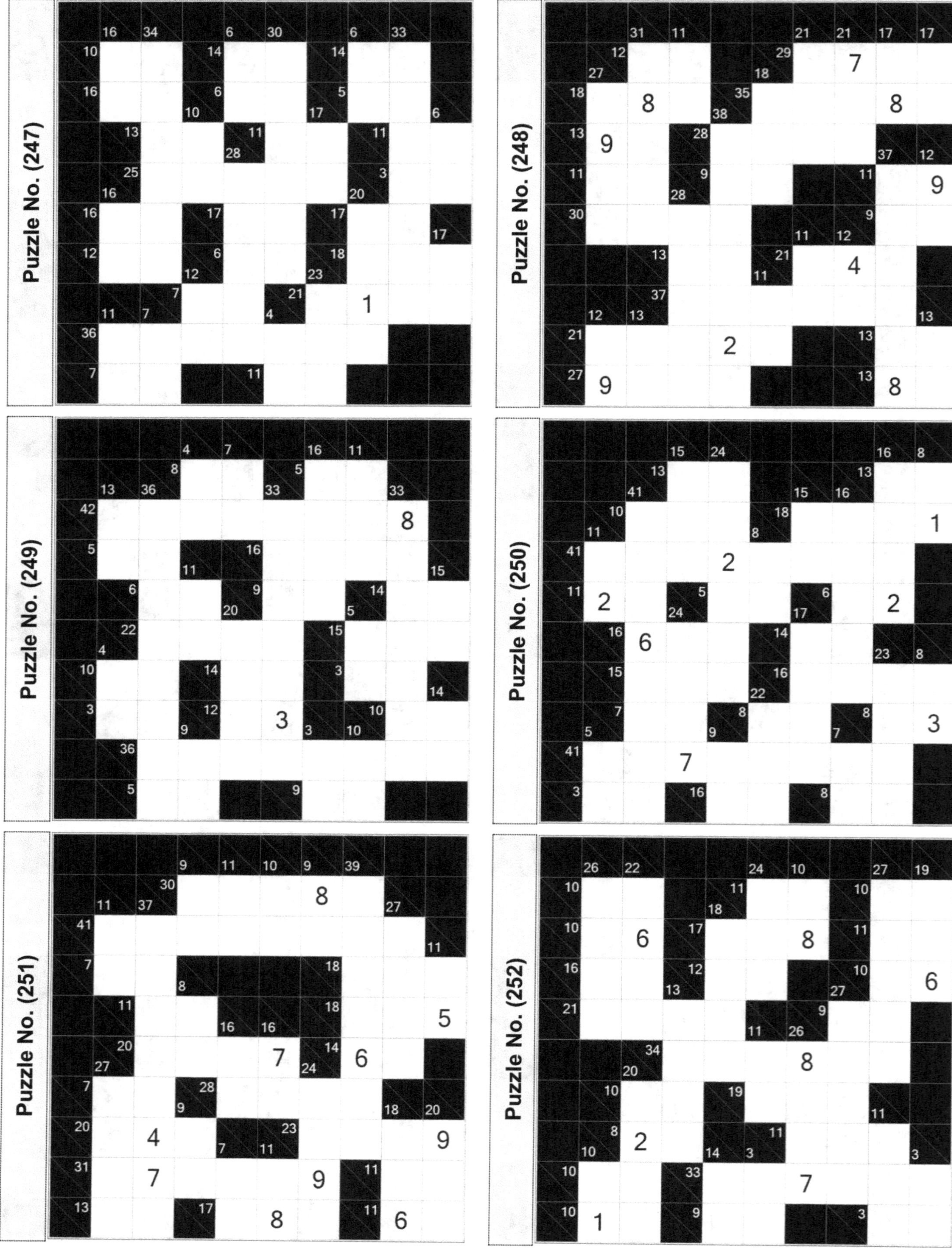

Solution on Page (182)

(44)

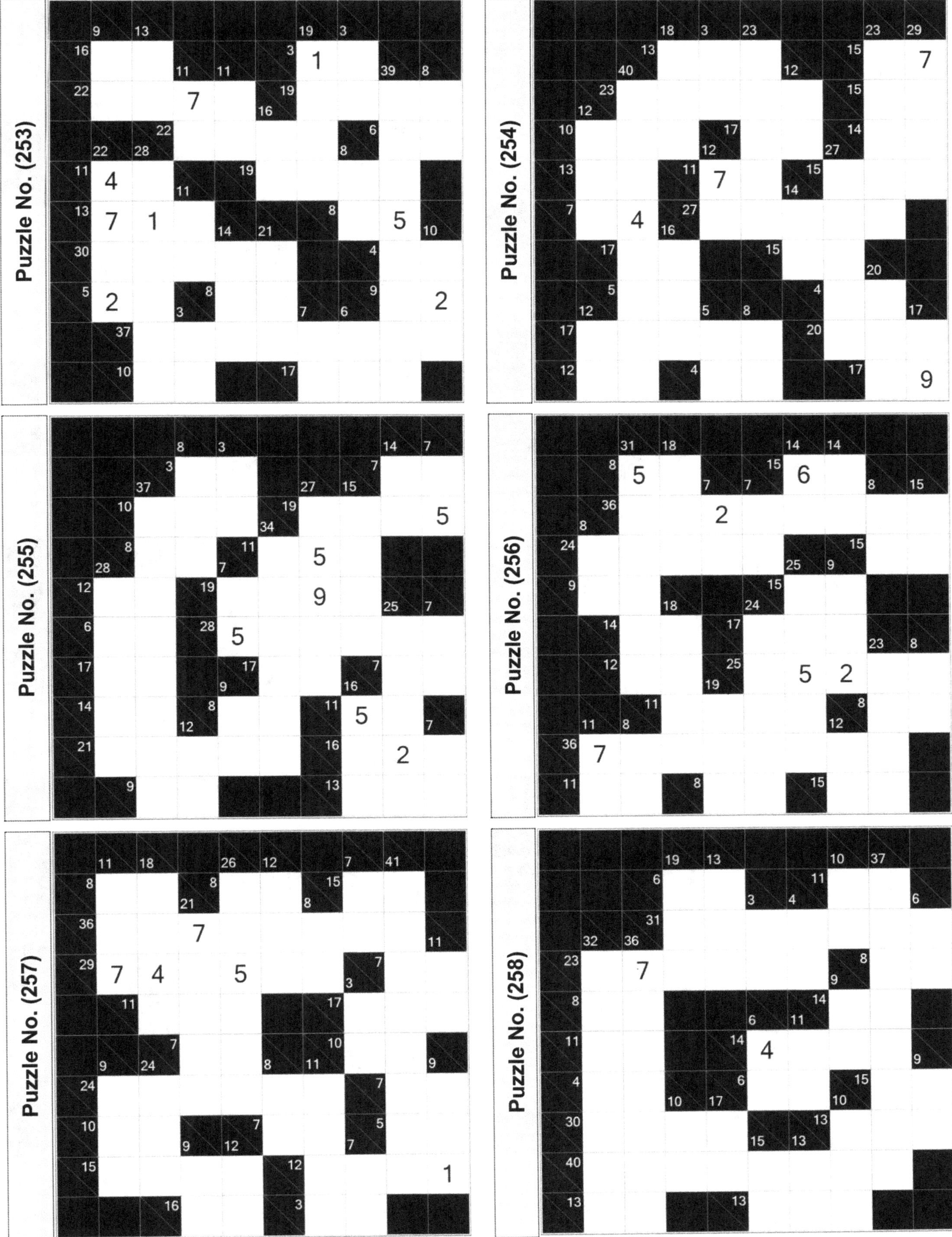

Solution on Page (182)

(45)

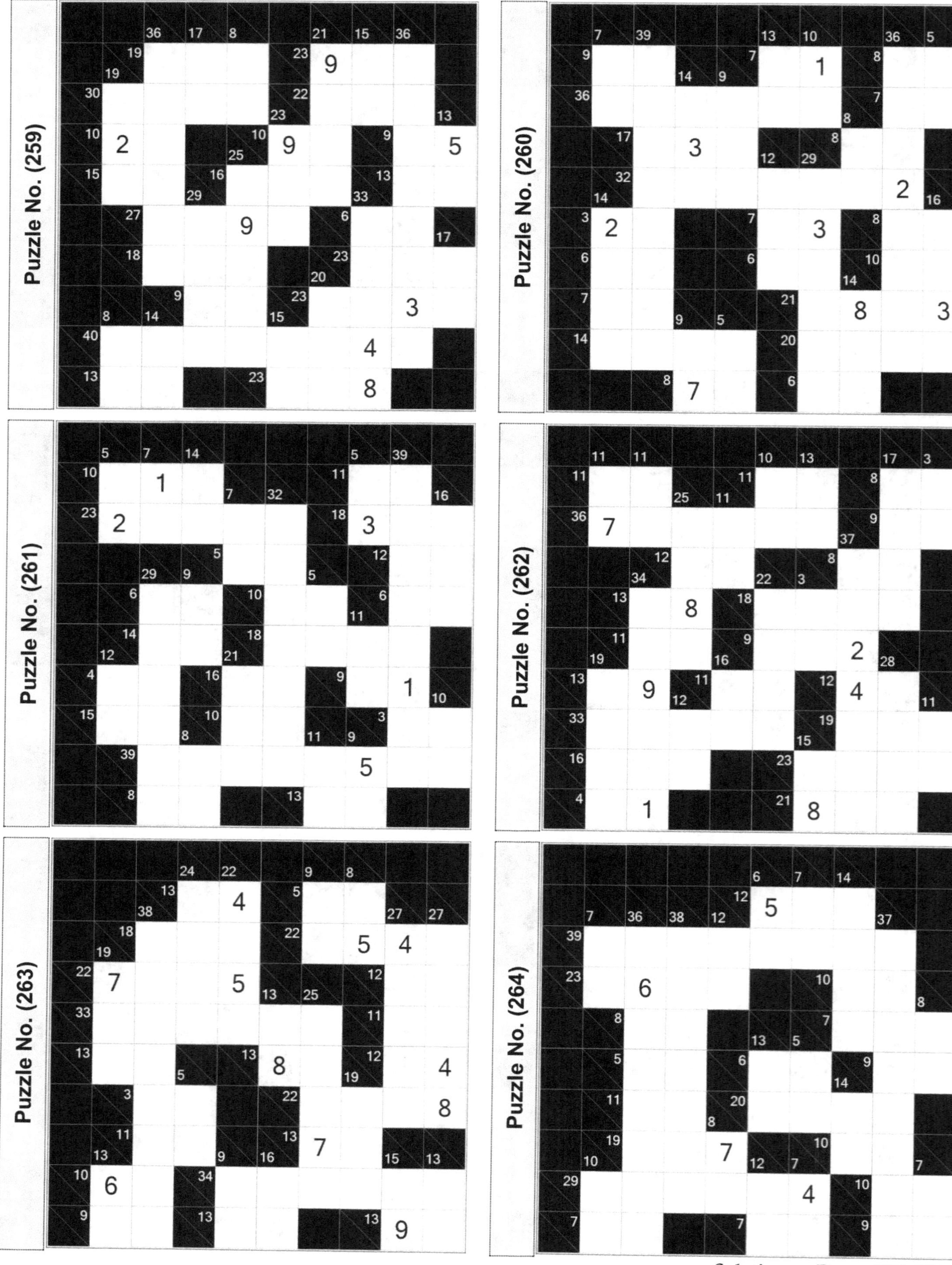

Puzzle No. (259)

Puzzle No. (260)

Puzzle No. (261)

Puzzle No. (262)

Puzzle No. (263)

Puzzle No. (264)

Solution on Pages (182-183)

(46)

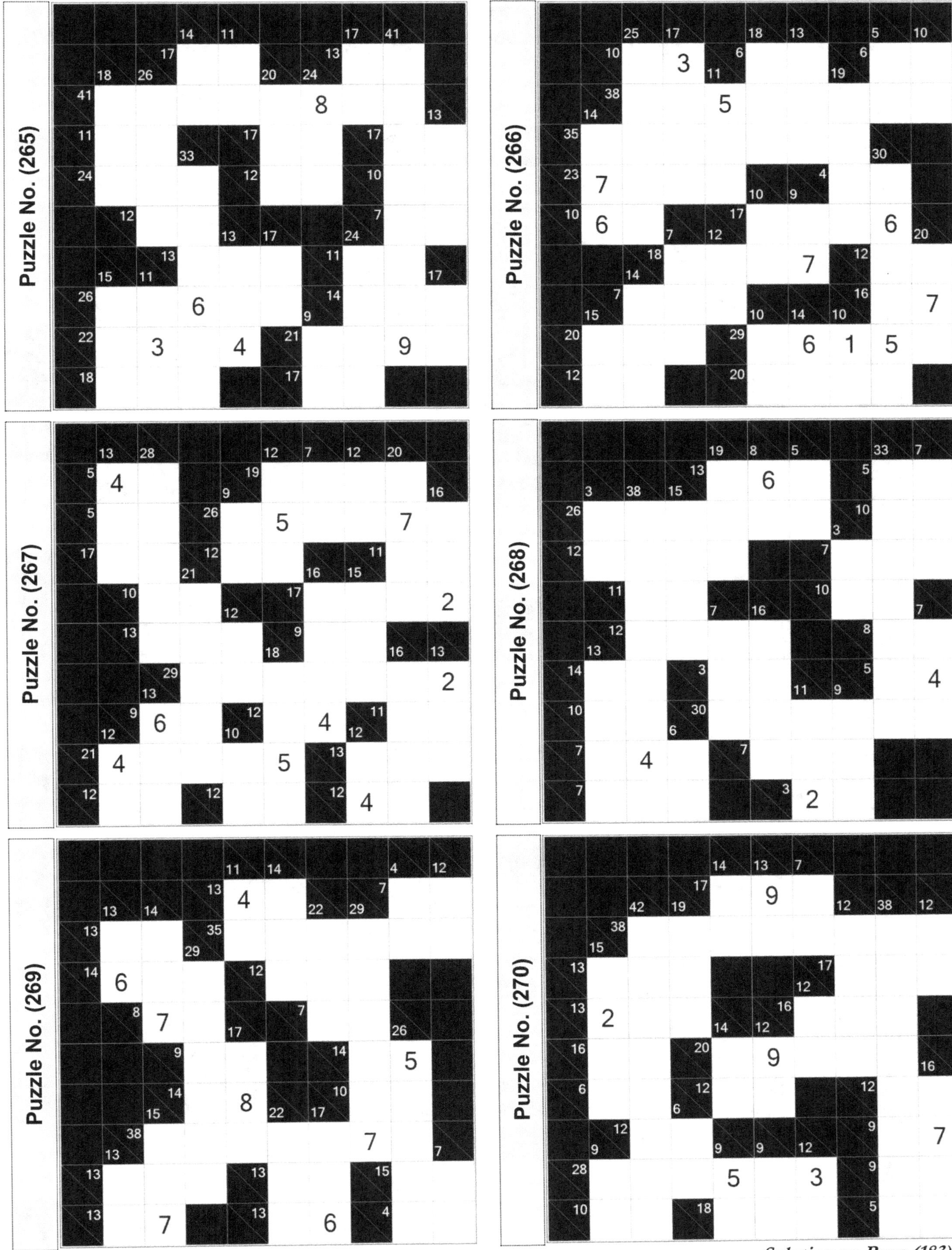

Puzzle No. (265)

Puzzle No. (266)

Puzzle No. (267)

Puzzle No. (268)

Puzzle No. (269)

Puzzle No. (270)

Solution on Page (183)

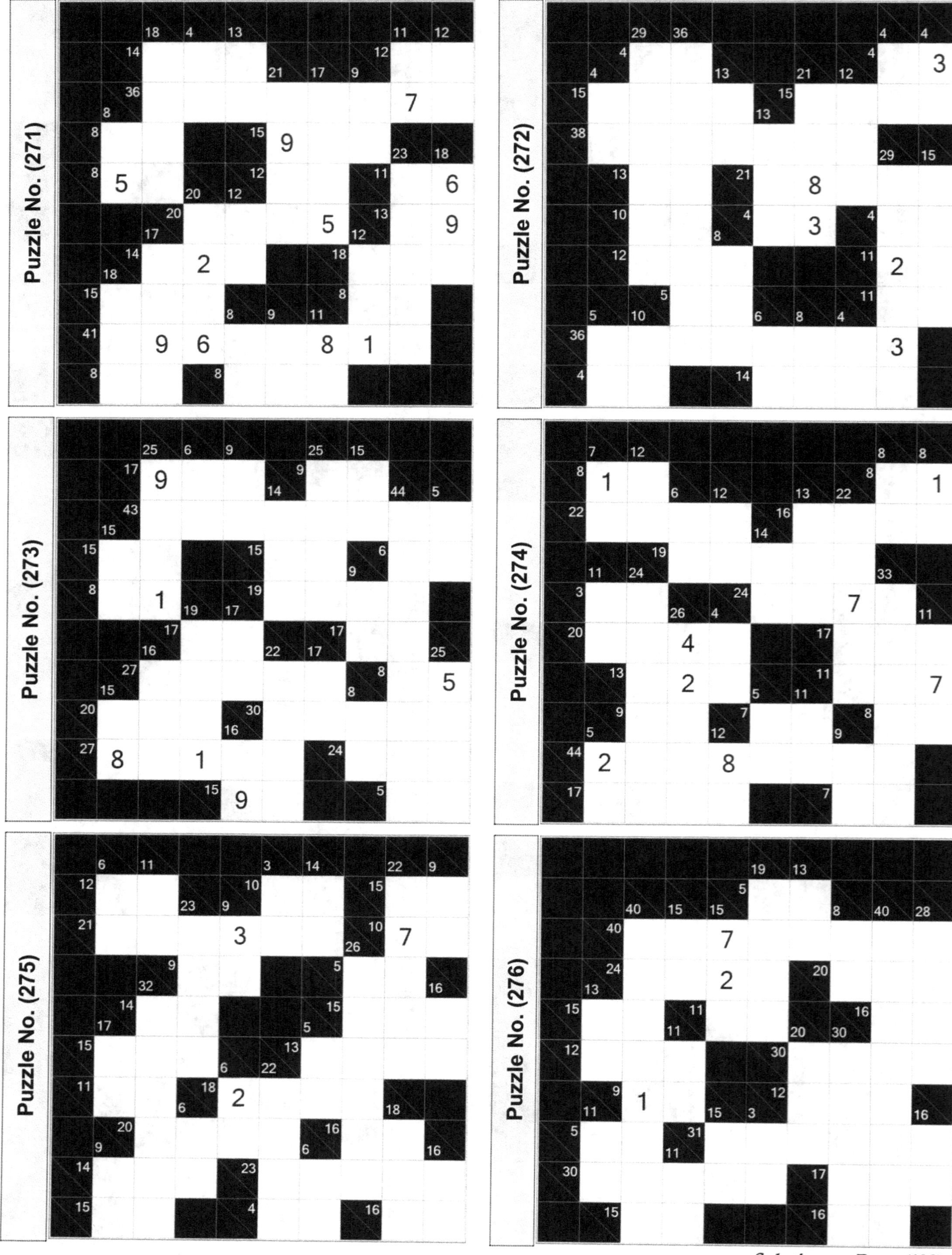

Solution on Page (183)

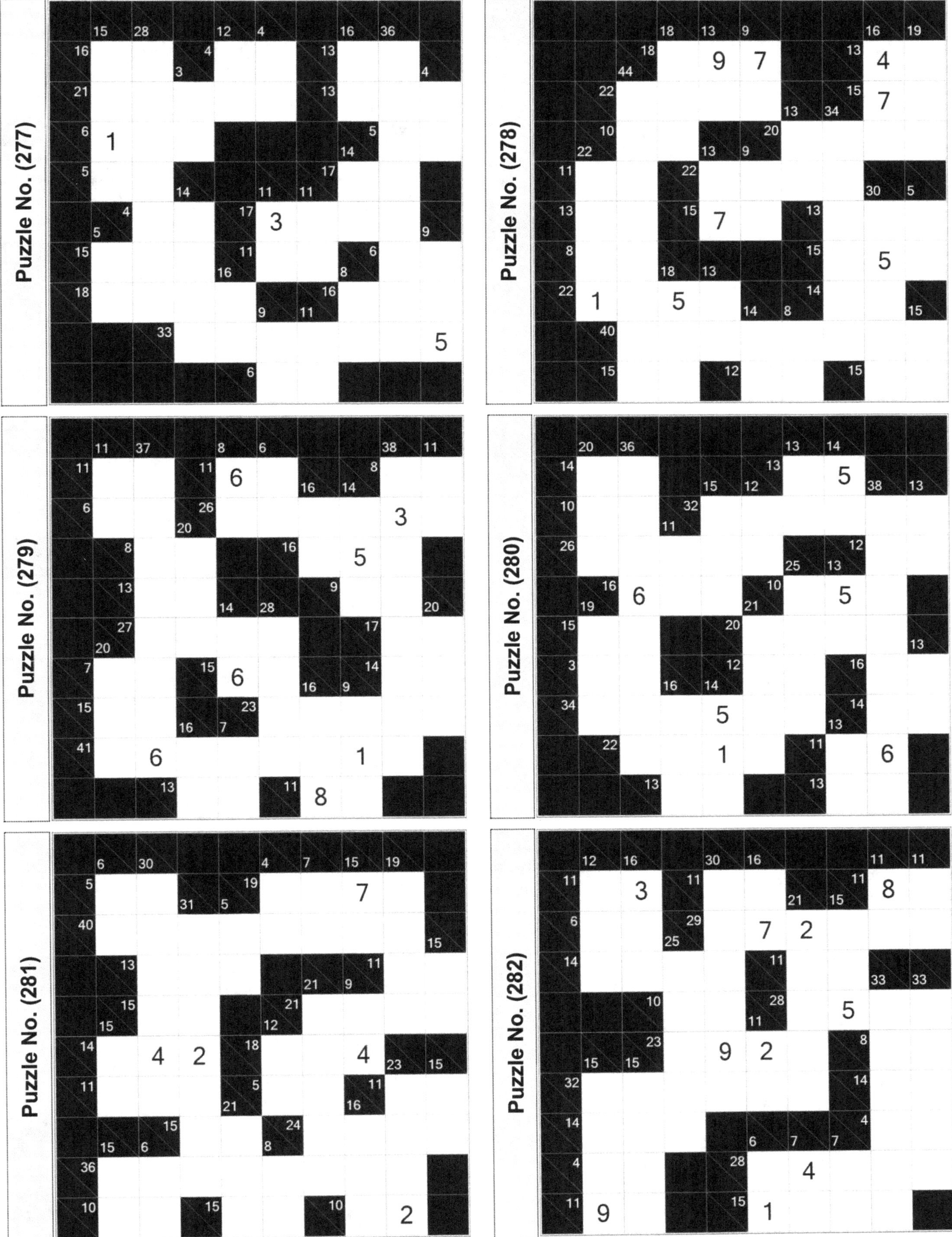

Puzzle No. (277)
Puzzle No. (278)
Puzzle No. (279)
Puzzle No. (280)
Puzzle No. (281)
Puzzle No. (282)

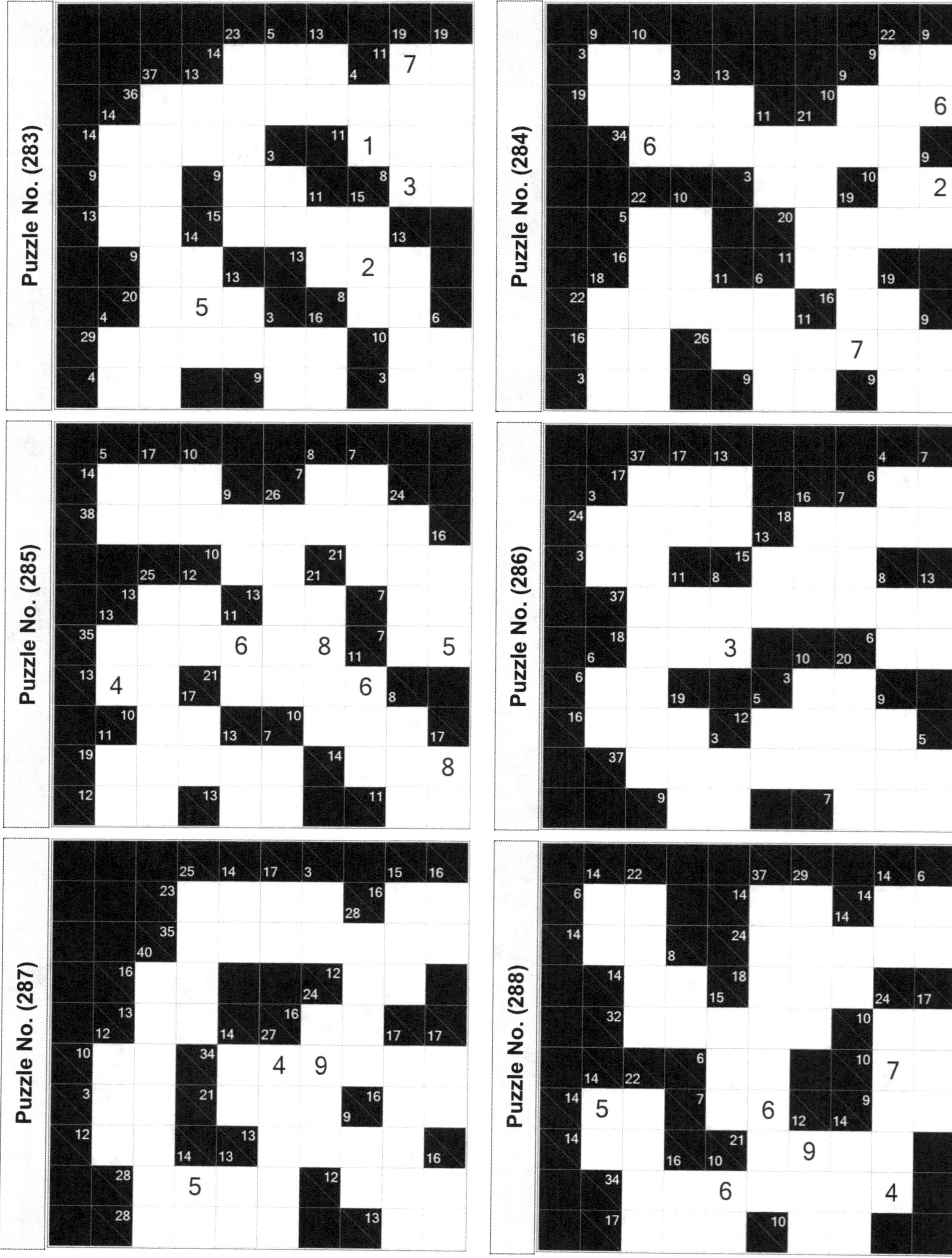

Puzzle No. (283)

Puzzle No. (284)

Puzzle No. (285)

Puzzle No. (286)

Puzzle No. (287)

Puzzle No. (288)

Solution on Page (184)

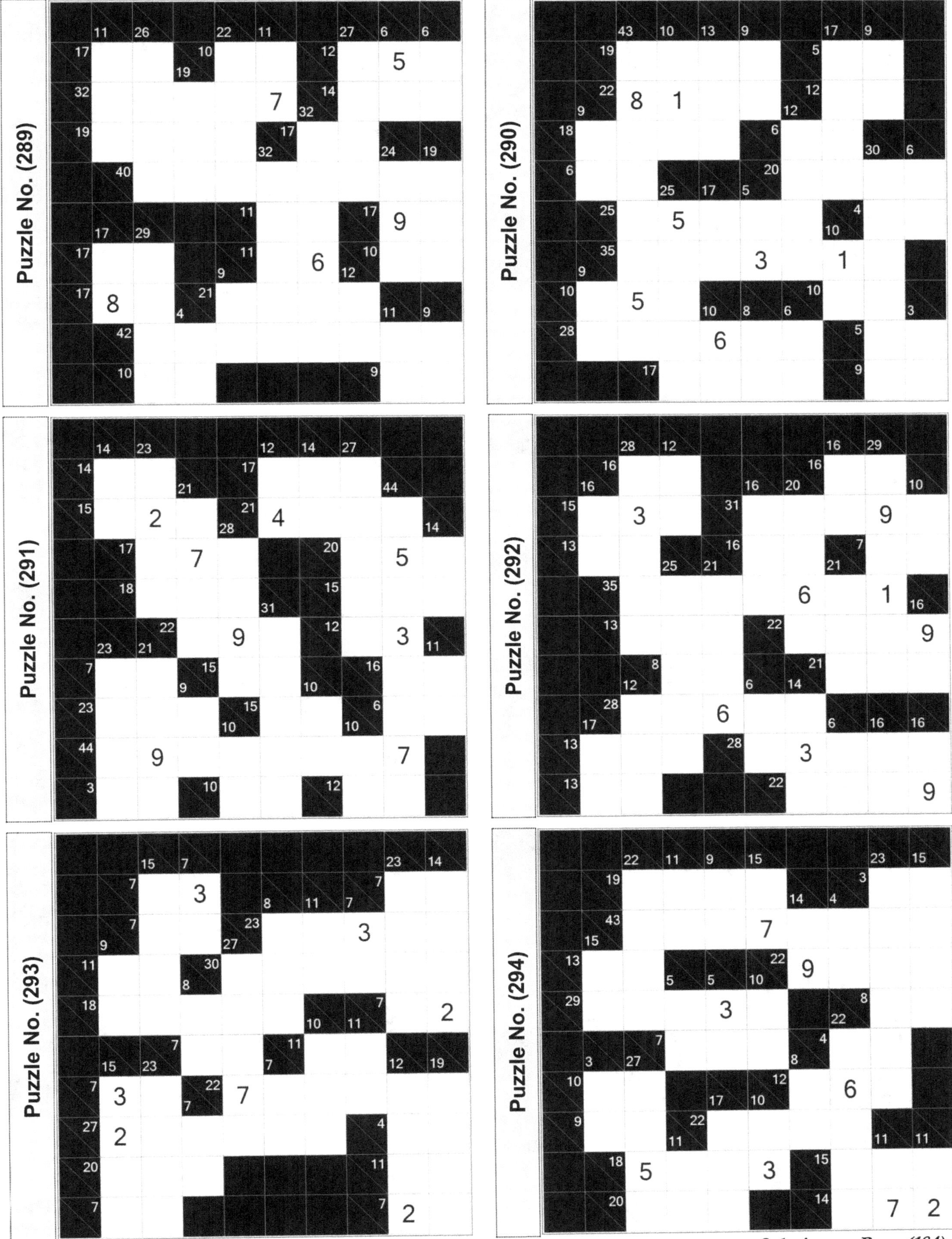

Puzzle No. (289)

Puzzle No. (290)

Puzzle No. (291)

Puzzle No. (292)

Puzzle No. (293)

Puzzle No. (294)

Solution on Page (184)

Solution on Page (184)

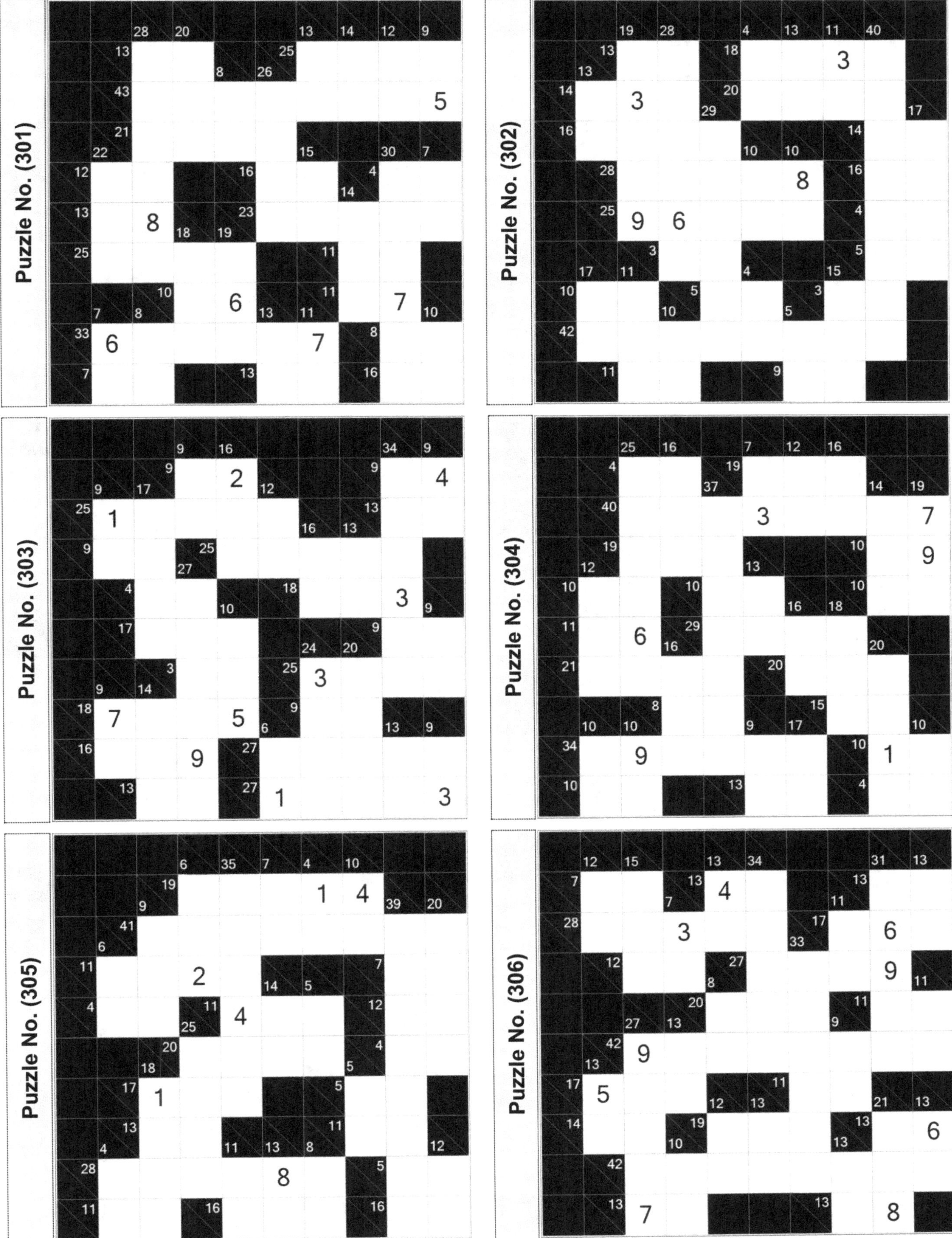

Puzzle No. (301)
Puzzle No. (302)
Puzzle No. (303)
Puzzle No. (304)
Puzzle No. (305)
Puzzle No. (306)
Solution on Page (185)

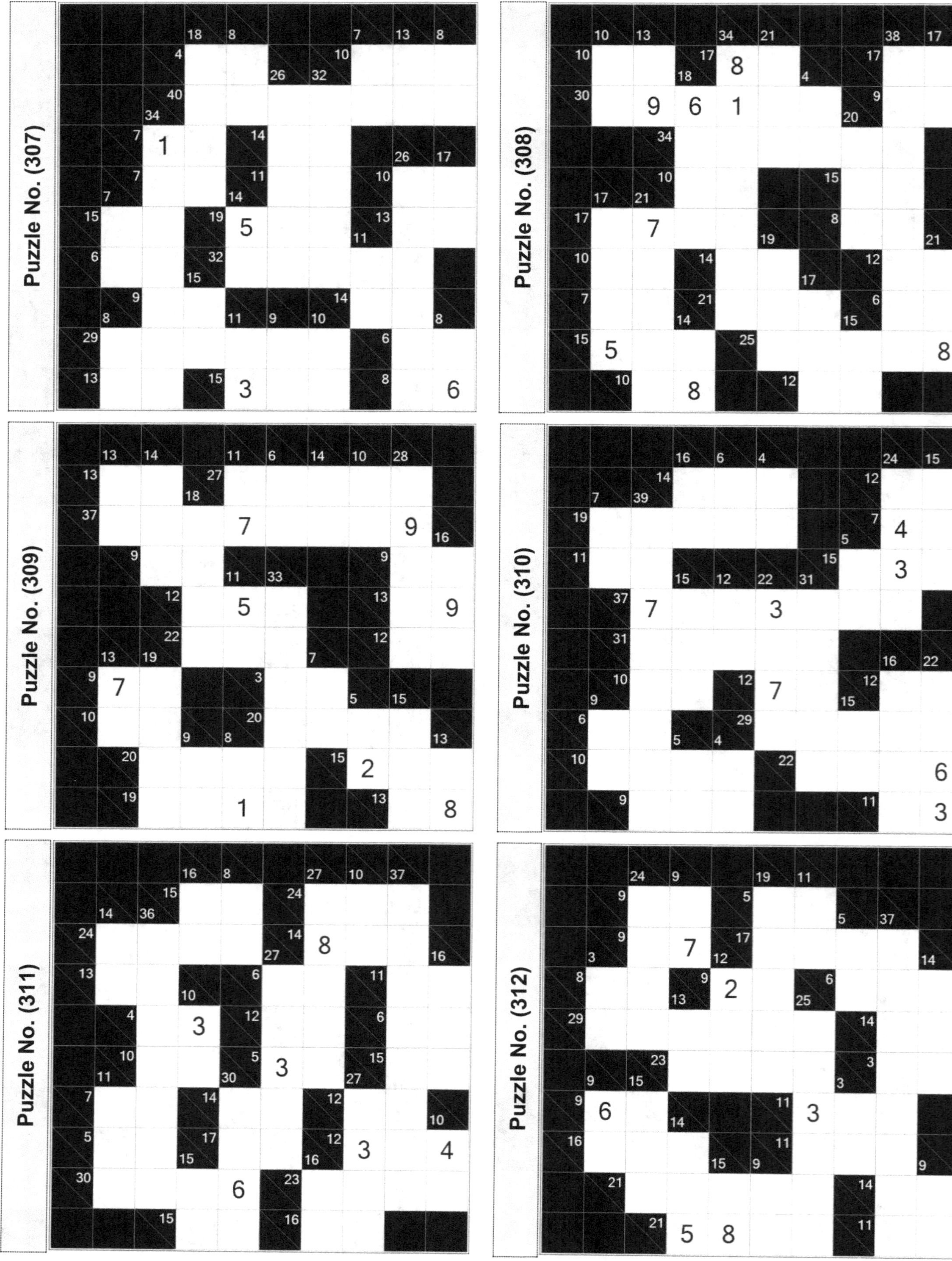

Puzzle No. (307)

Puzzle No. (308)

Puzzle No. (309)

Puzzle No. (310)

Puzzle No. (311)

Puzzle No. (312)

Solution on Page (185)

(54)

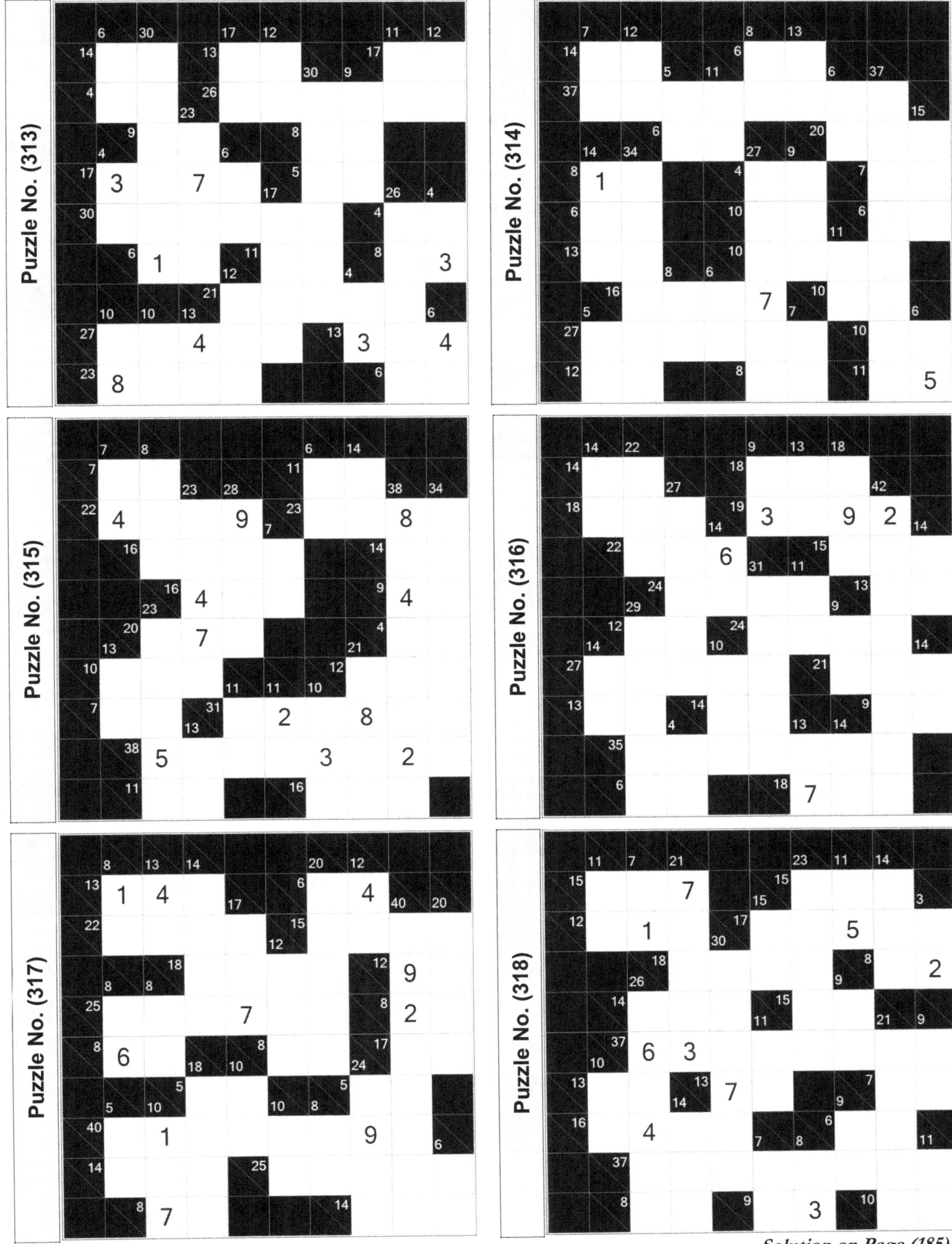

Puzzle No. (313)

Puzzle No. (314)

Puzzle No. (315)

Puzzle No. (316)

Puzzle No. (317)

Puzzle No. (318)

Solution on Page (185)

(55)

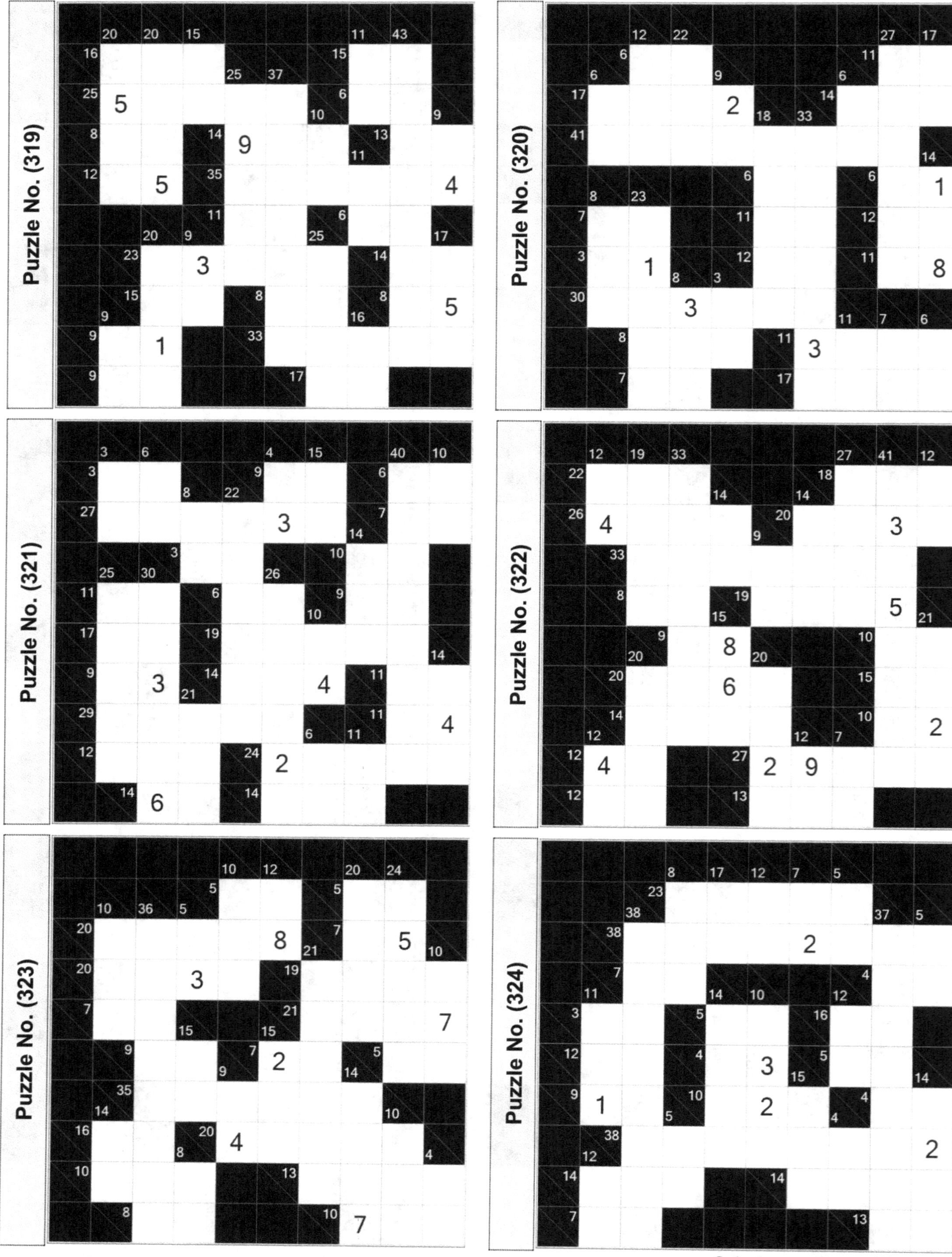

Puzzle No. (319)

Puzzle No. (320)

Puzzle No. (321)

Puzzle No. (322)

Puzzle No. (323)

Puzzle No. (324)

Solution on Pages (185-186)

(56)

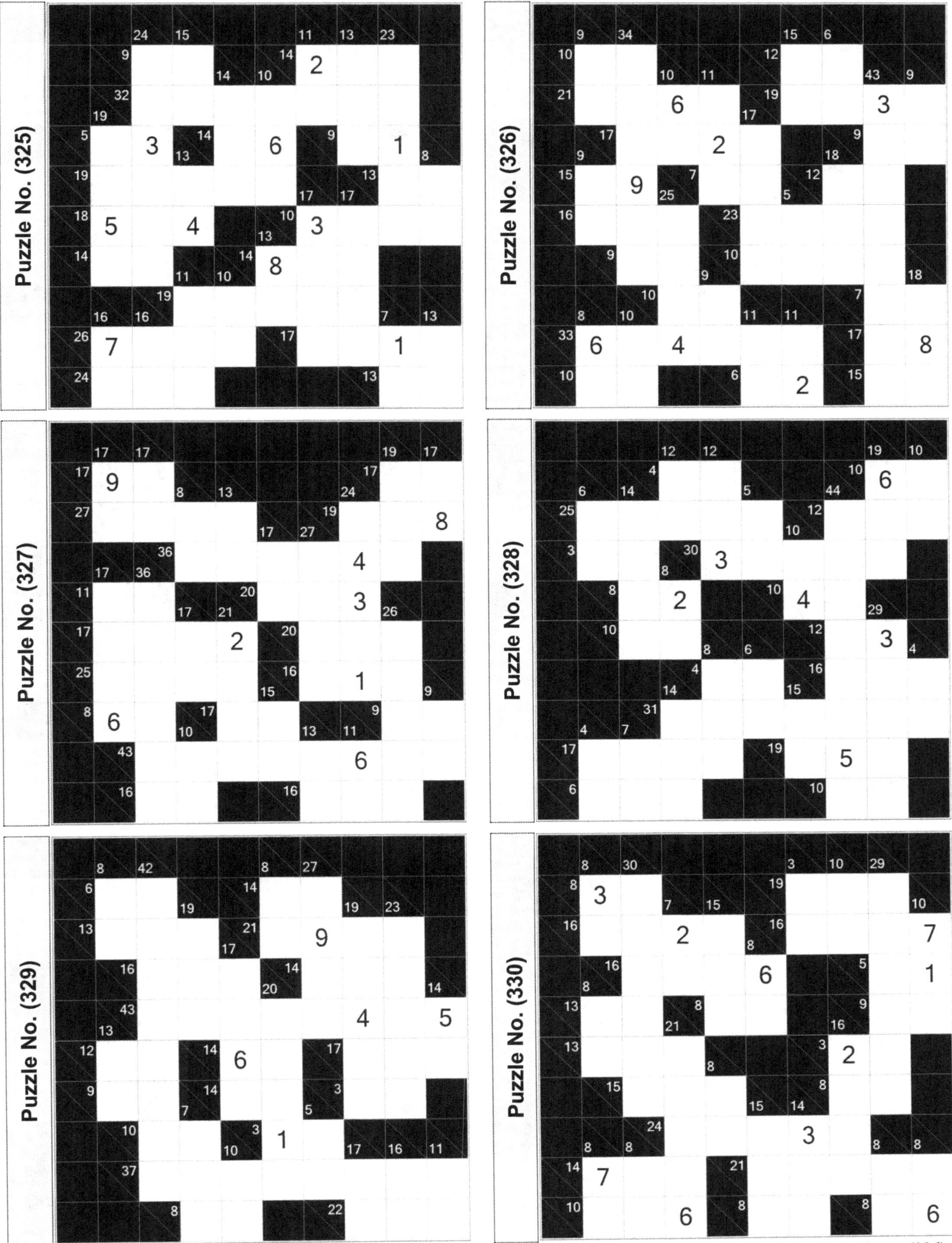

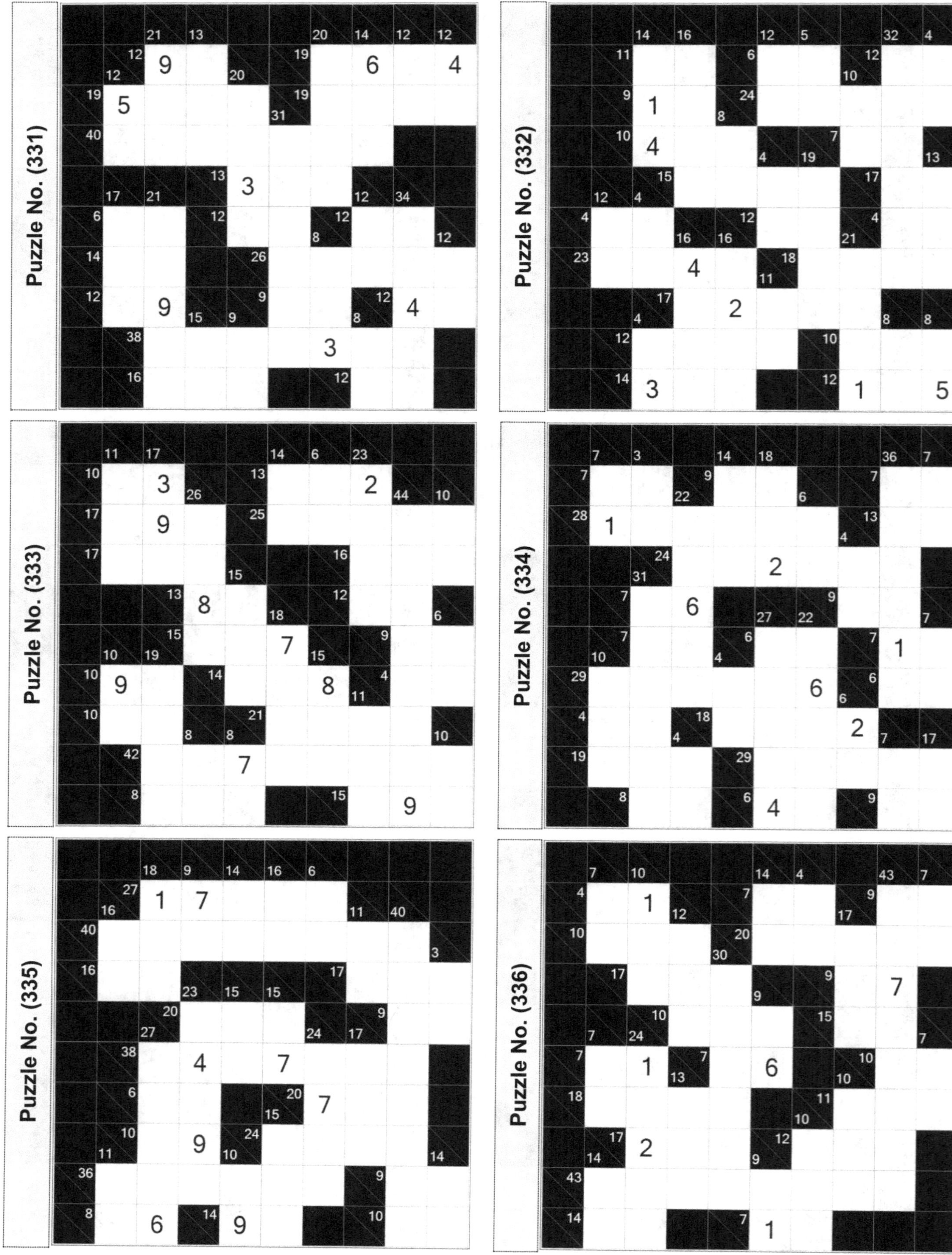

Puzzle No. (331)

Puzzle No. (332)

Puzzle No. (333)

Puzzle No. (334)

Puzzle No. (335)

Puzzle No. (336)

Solution on Page (186)

(58)

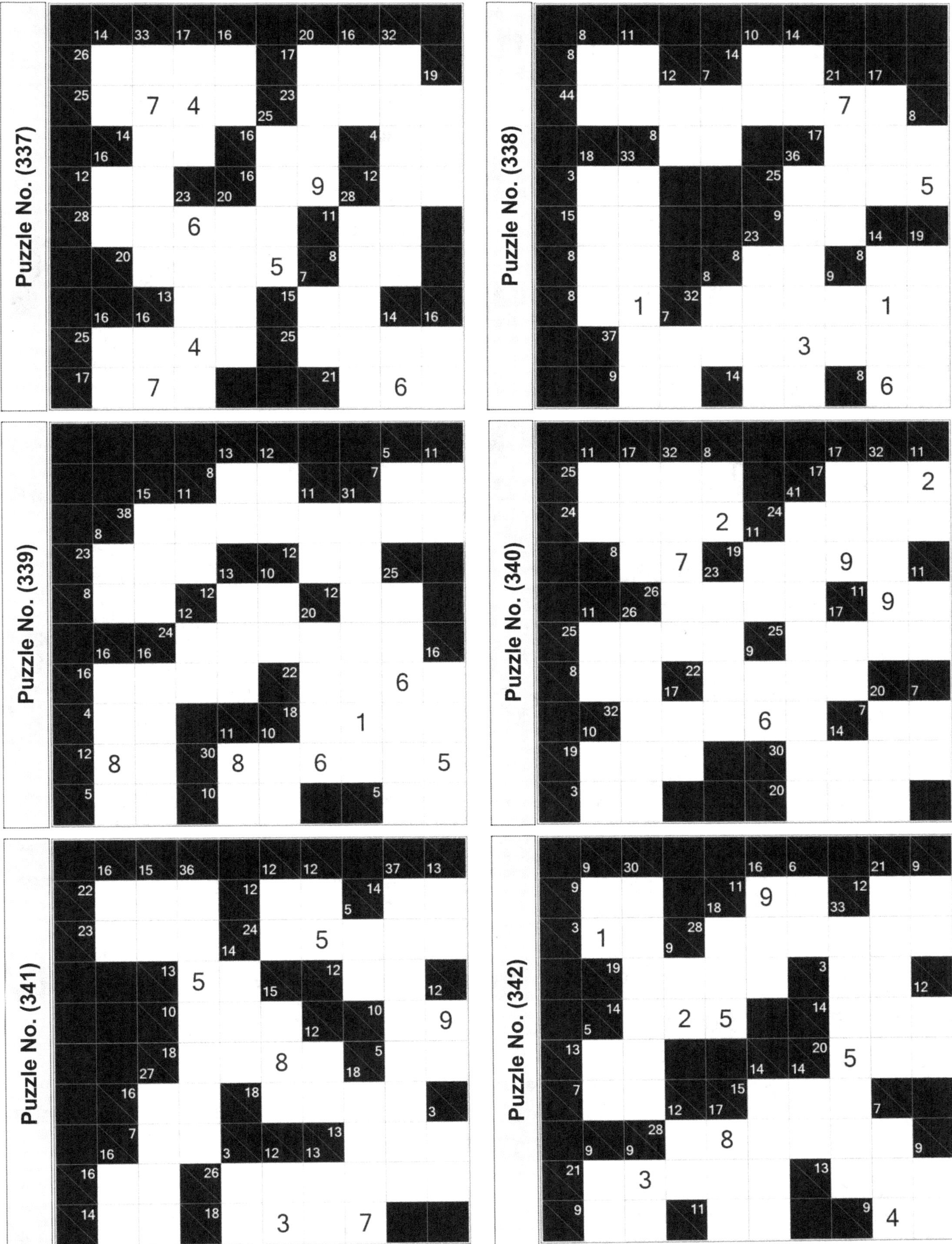

Solution on Pages (186-187)

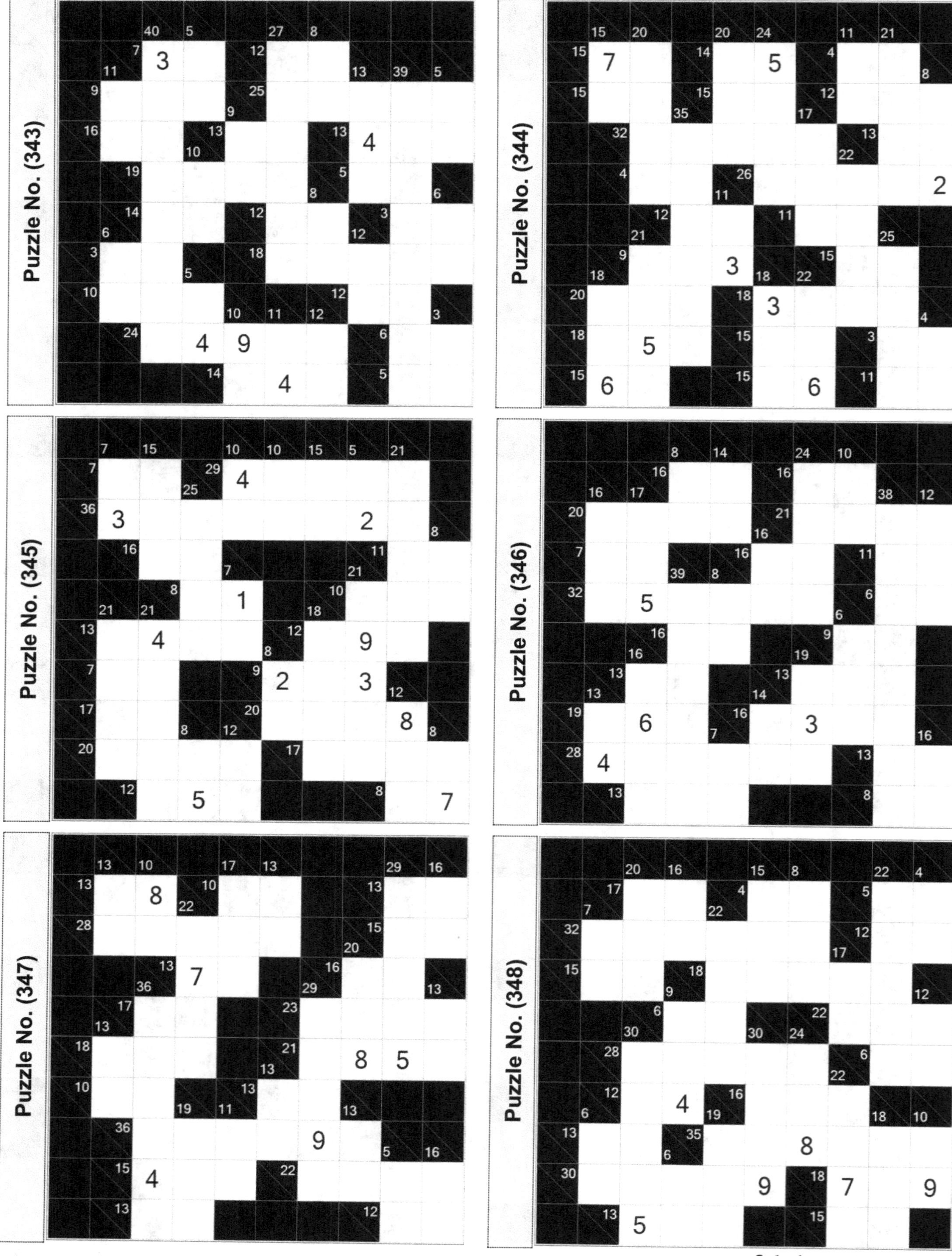

Puzzle No. (343)

Puzzle No. (344)

Puzzle No. (345)

Puzzle No. (346)

Puzzle No. (347)

Puzzle No. (348)

Solution on Page (187)

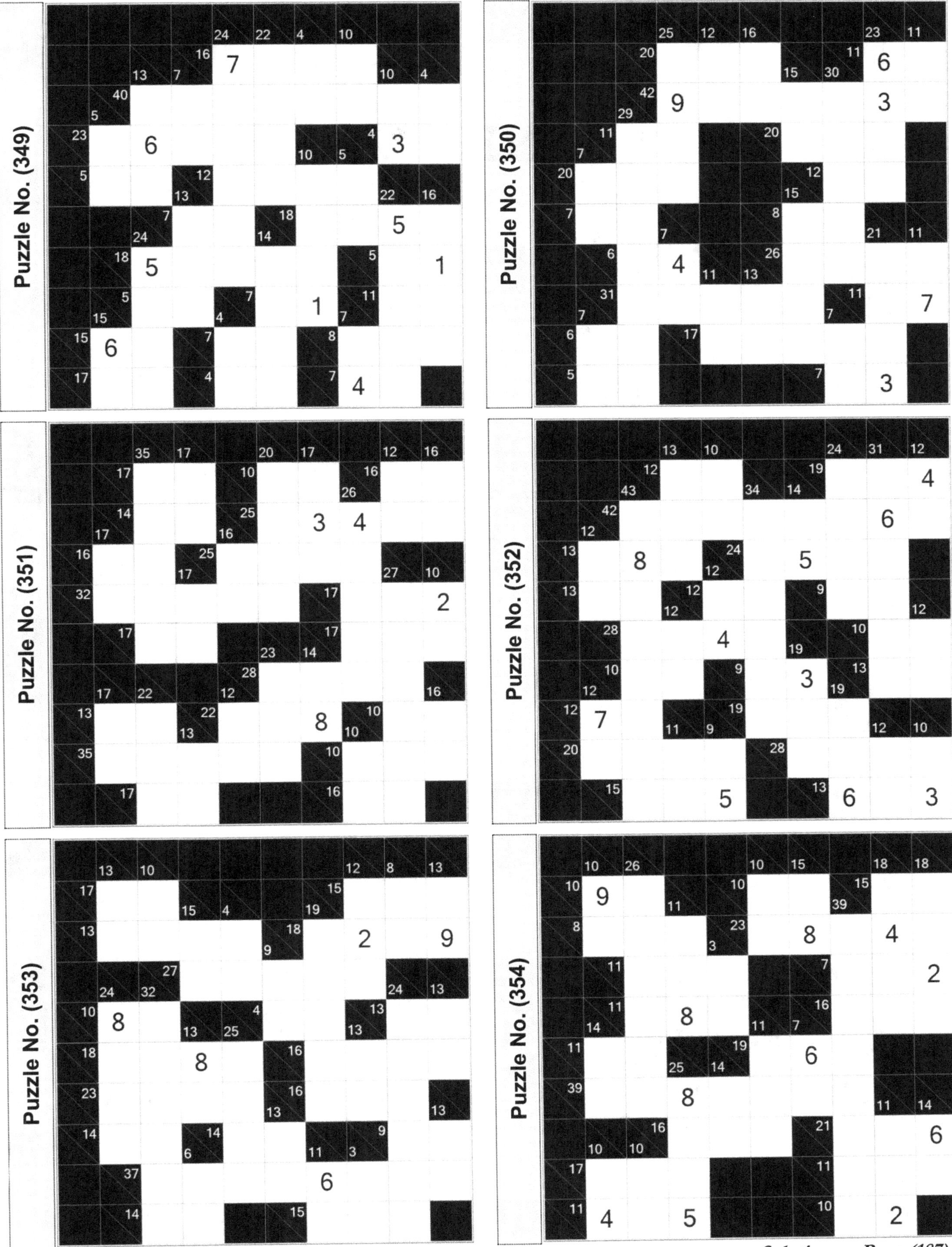

Puzzle No. (349)

Puzzle No. (350)

Puzzle No. (351)

Puzzle No. (352)

Puzzle No. (353)

Puzzle No. (354)

Solution on Page (187)

(61)

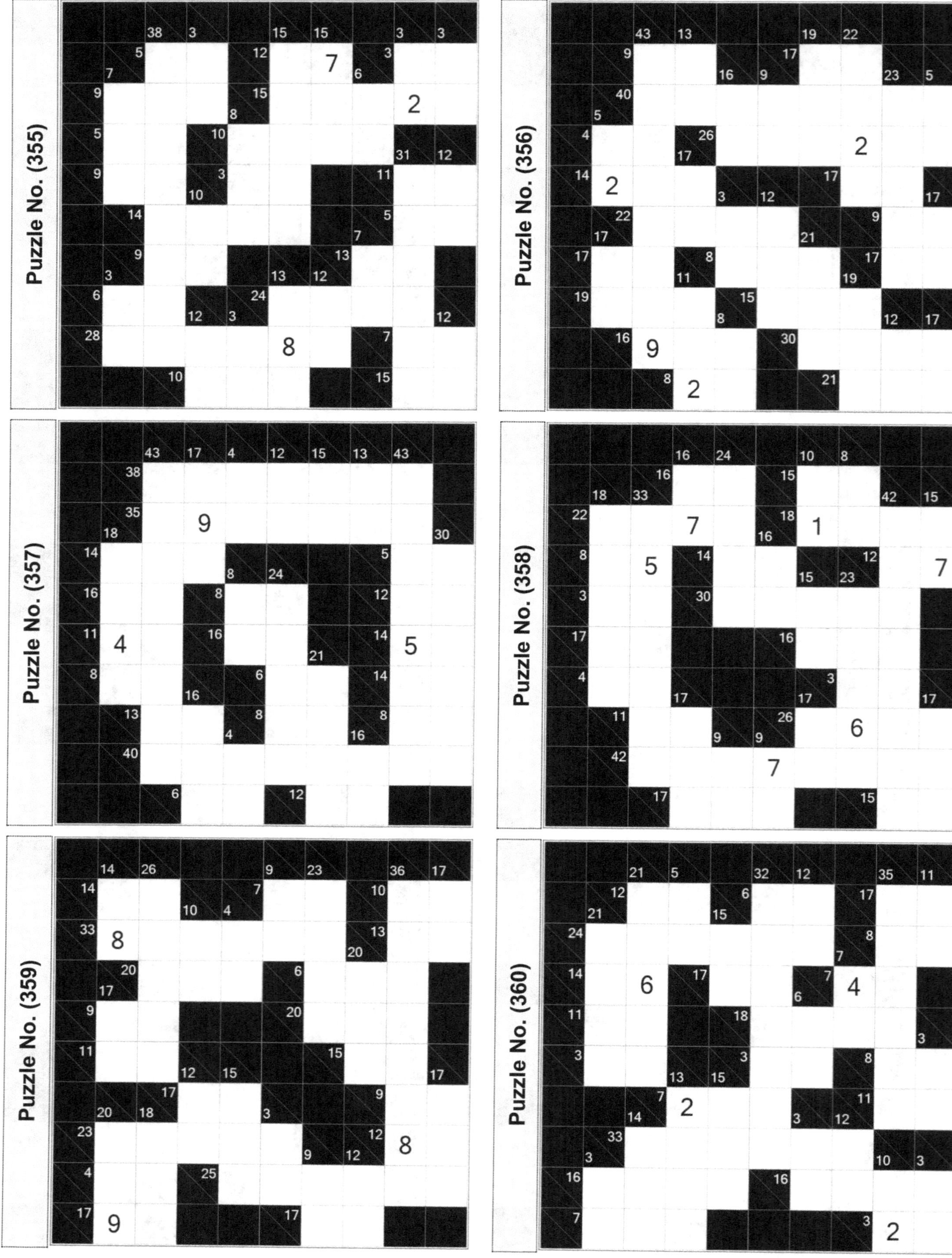

Puzzle No. (355)

Puzzle No. (356)

Puzzle No. (357)

Puzzle No. (358)

Puzzle No. (359)

Puzzle No. (360)

Solution on Page (187)

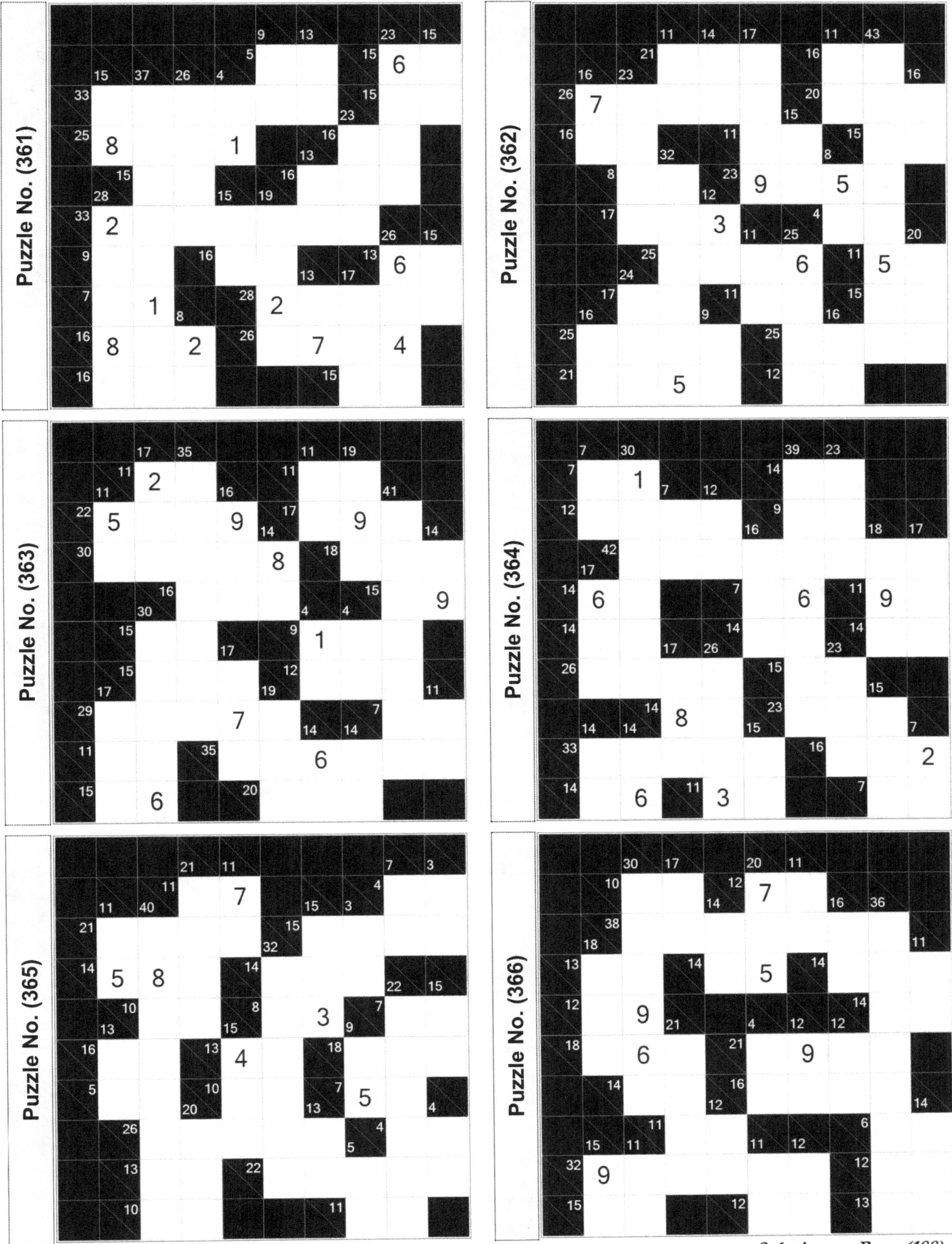

Puzzle No. (361)

Puzzle No. (362)

Puzzle No. (363)

Puzzle No. (364)

Puzzle No. (365)

Puzzle No. (366)

Solution on Page (188)

(63)

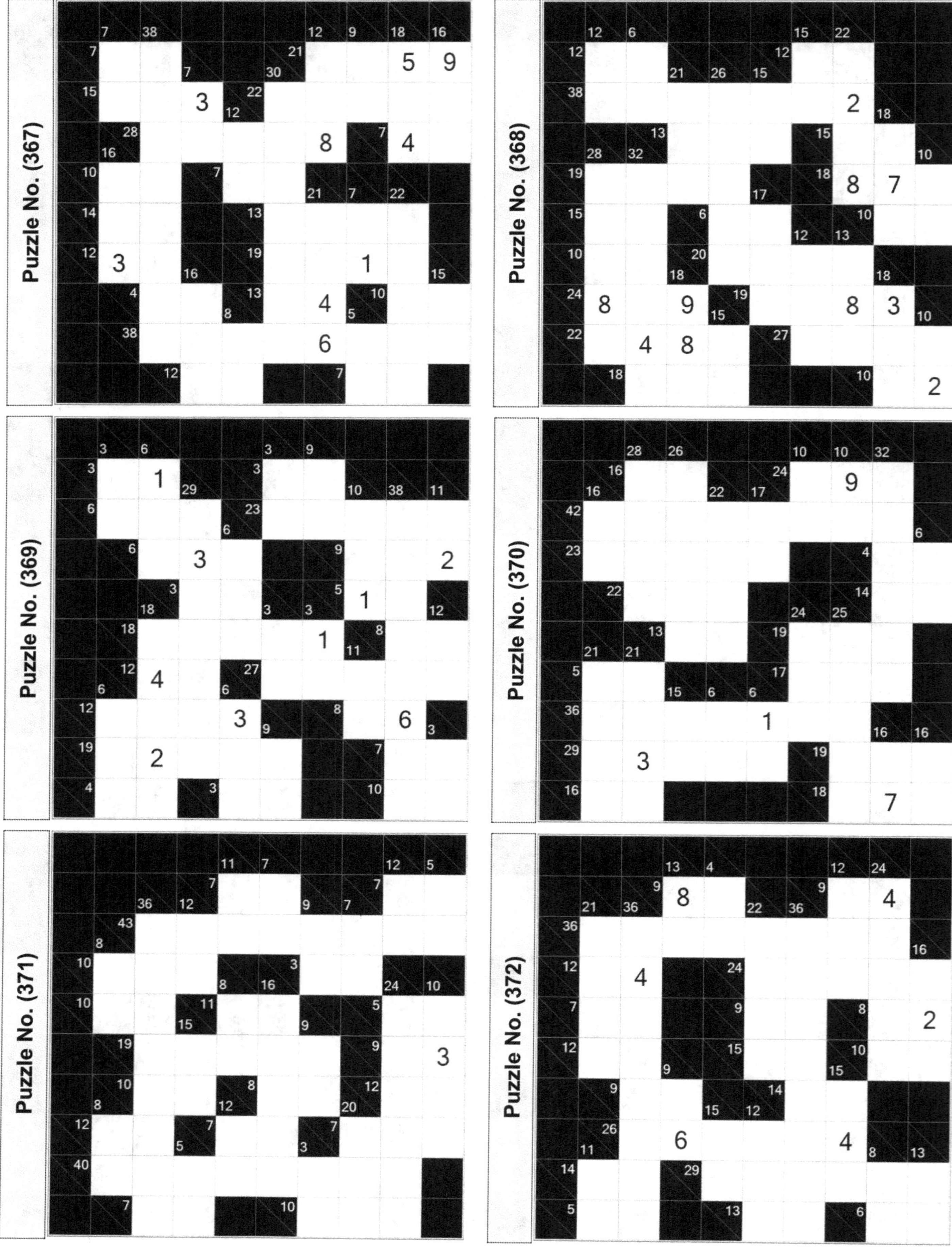

Solution on Page (188)

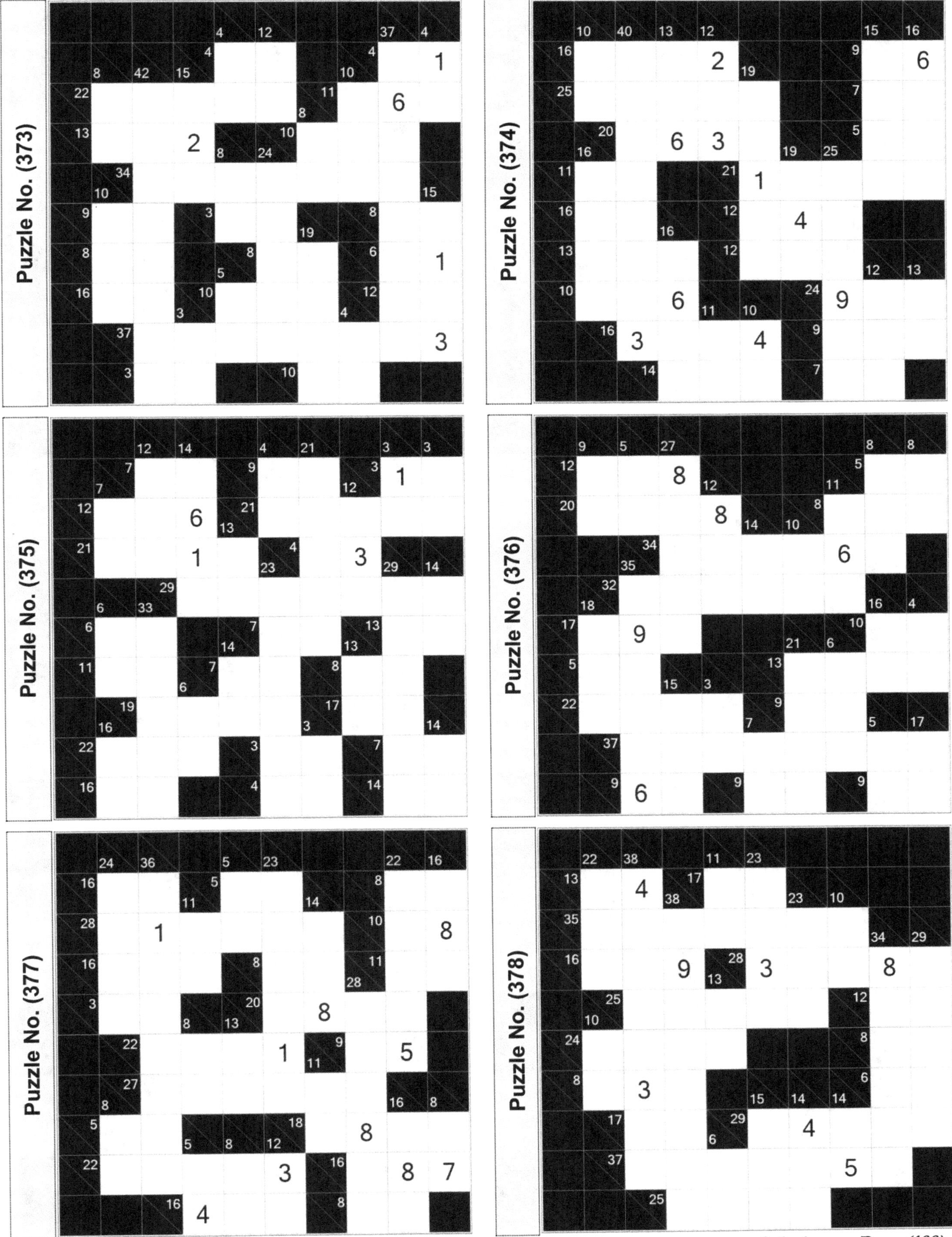

Solution on Page (188)

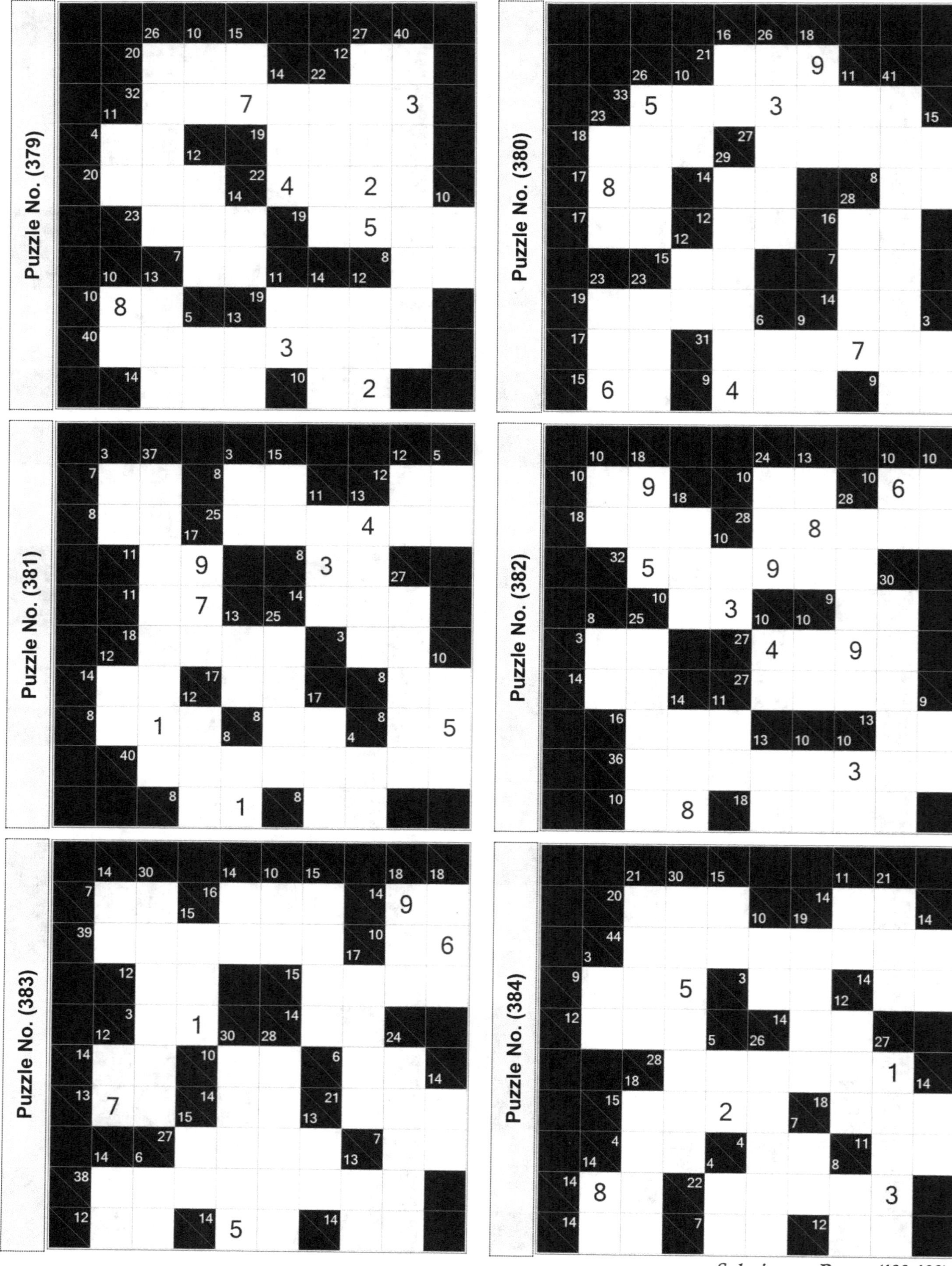

Solution on Pages (188-189)

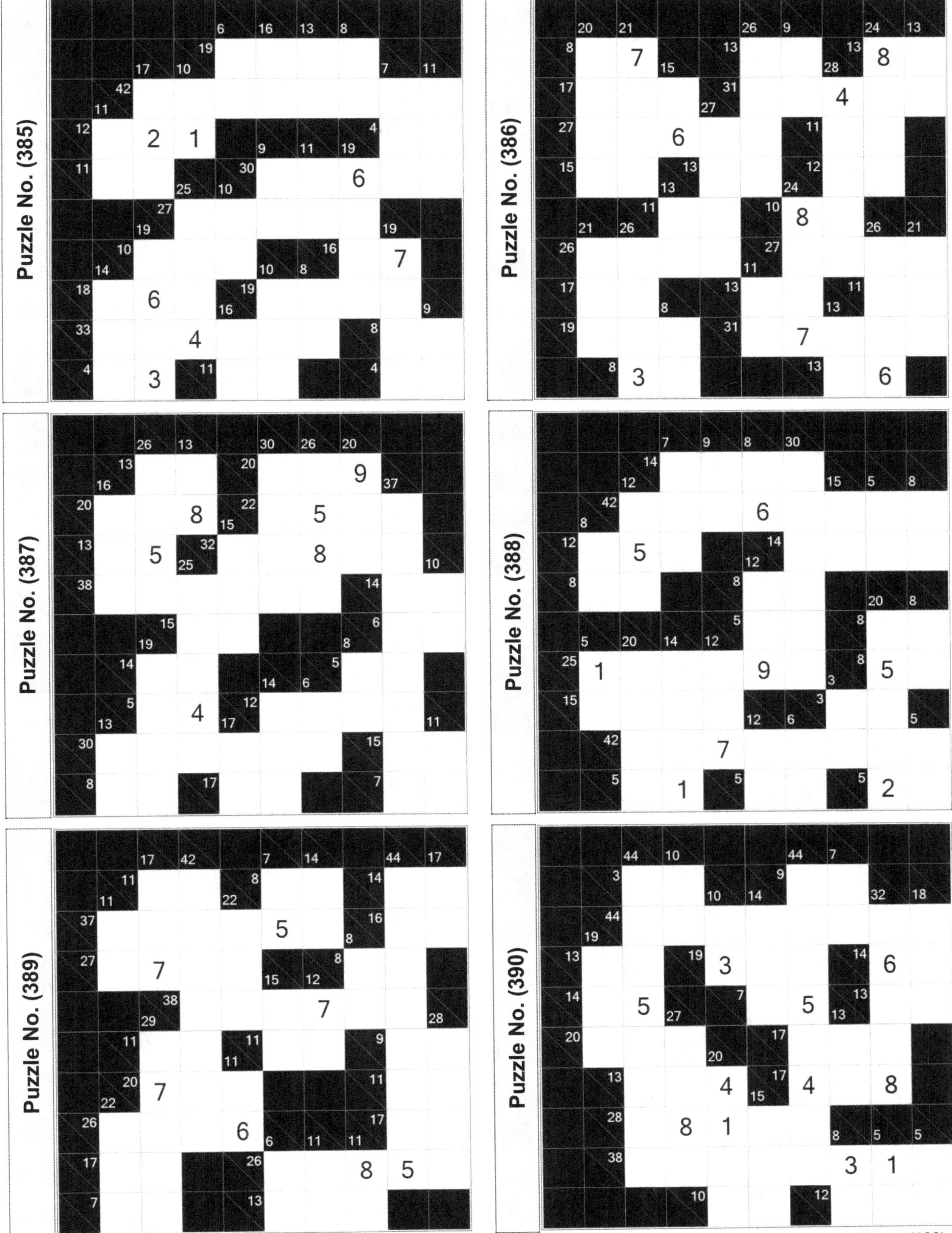

Puzzle No. (385)

Puzzle No. (386)

Puzzle No. (387)

Puzzle No. (388)

Puzzle No. (389)

Puzzle No. (390)

Solution on Page (189)

(67)

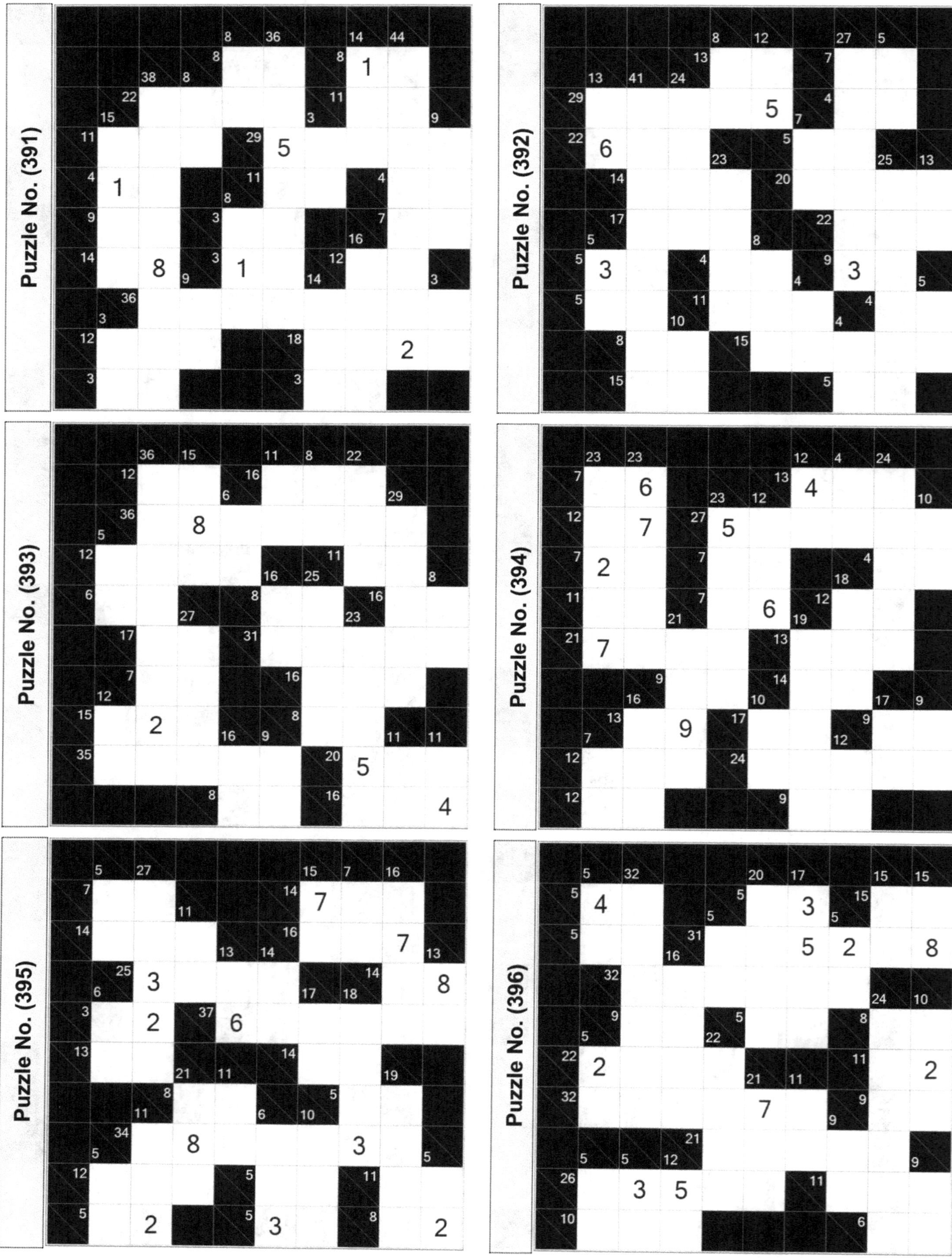

Puzzle No. (391)
Puzzle No. (392)
Puzzle No. (393)
Puzzle No. (394)
Puzzle No. (395)
Puzzle No. (396)
Solution on Page (189)

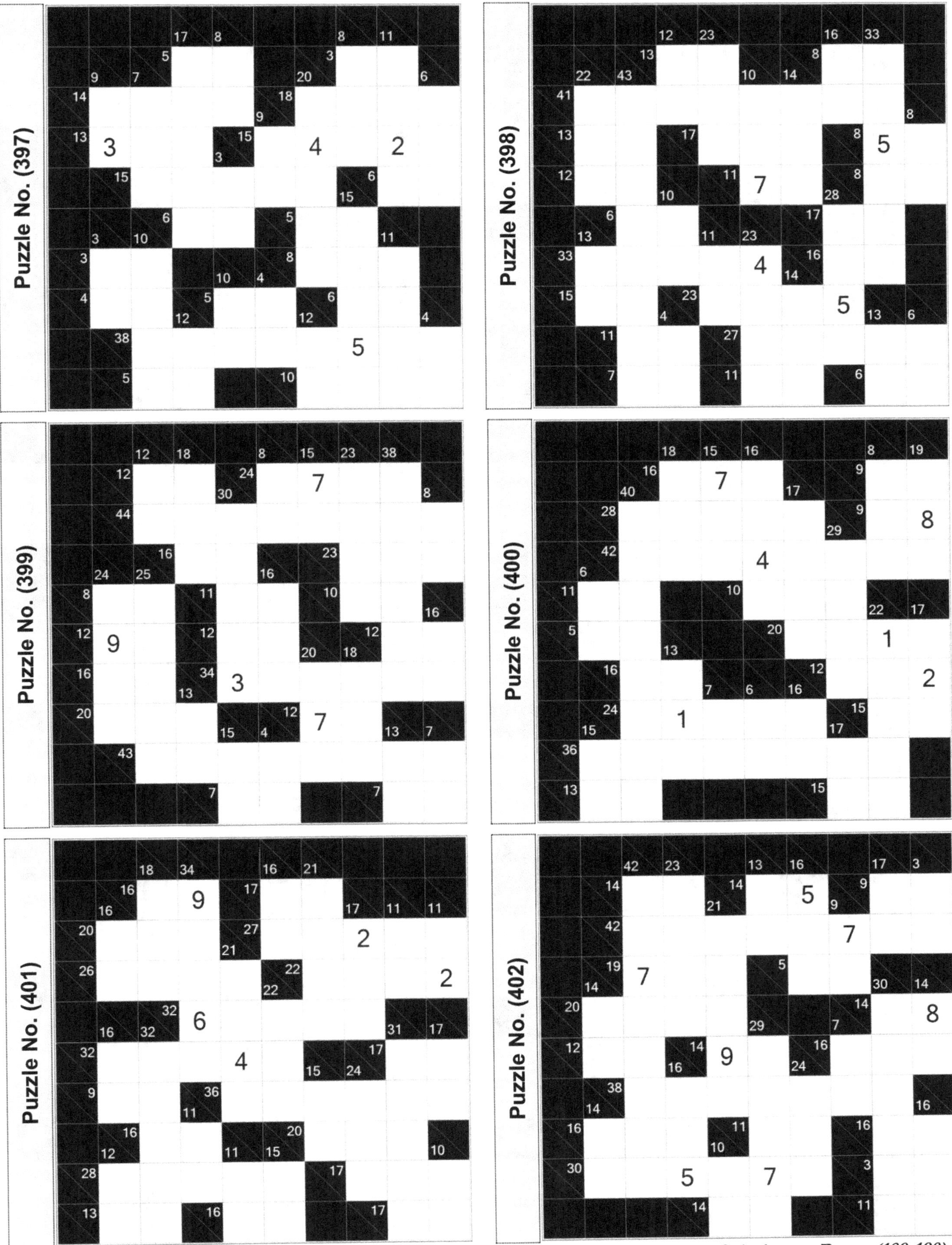

Solution on Pages (189-190)

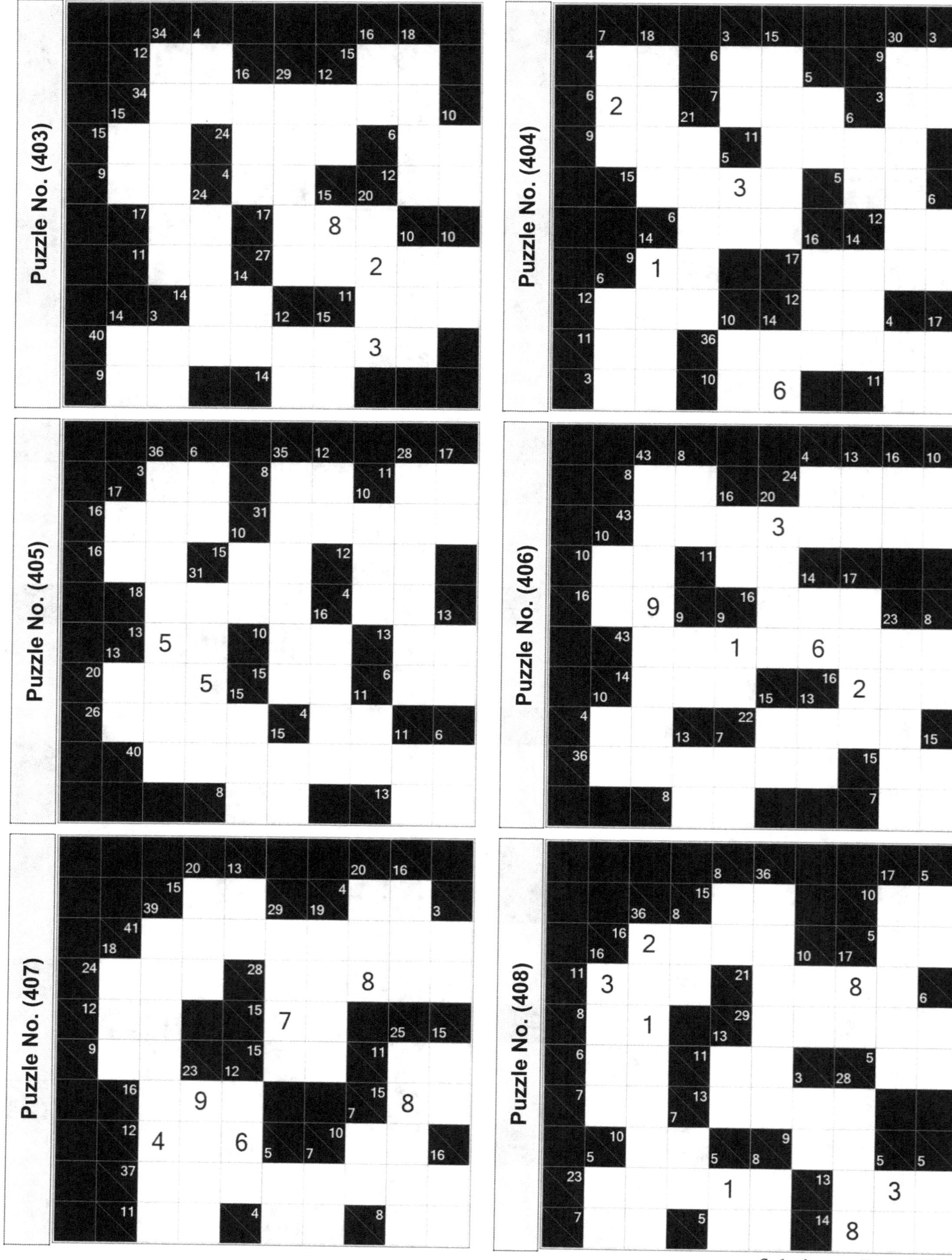

Solution on Page (190)

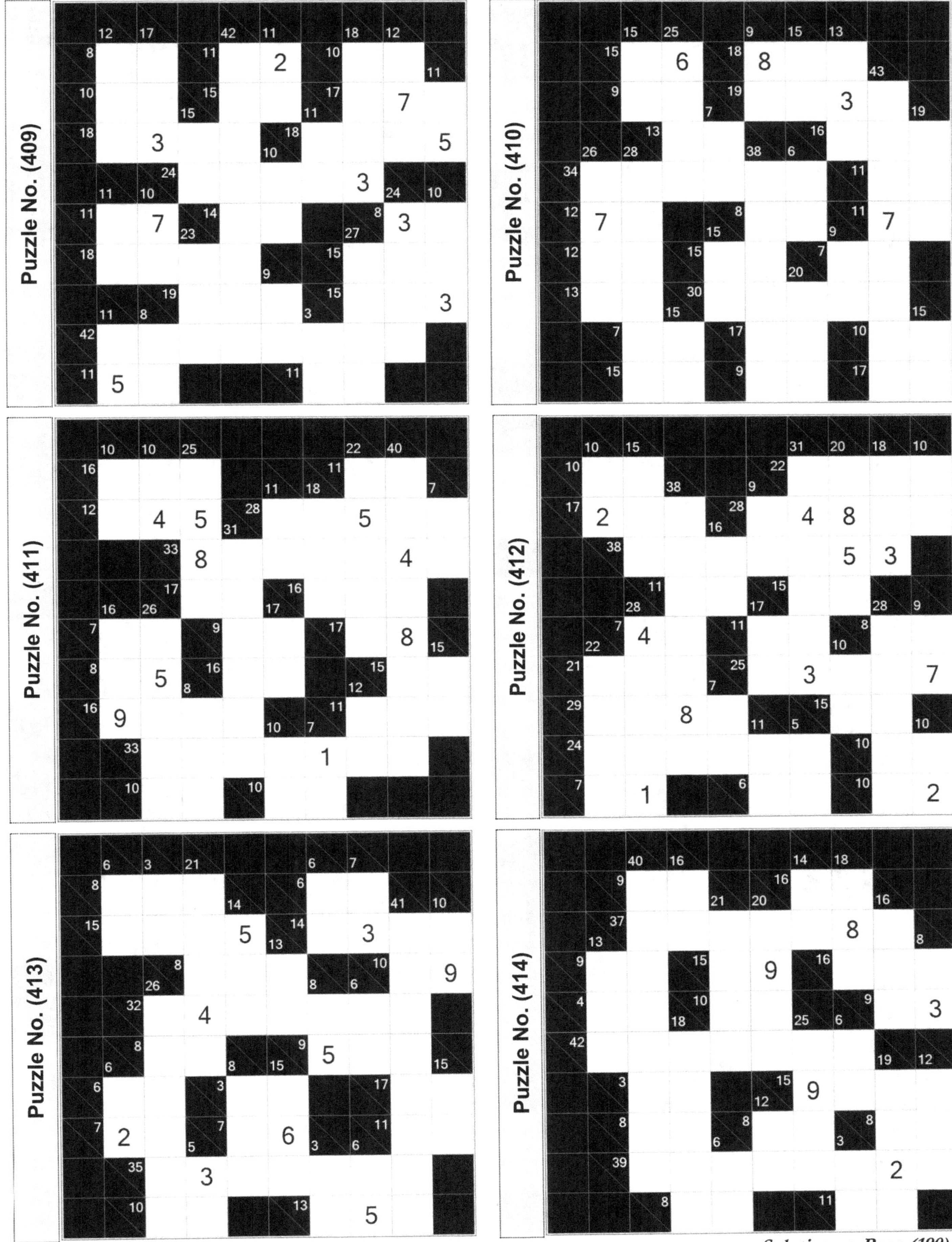

Solution on Page (190)

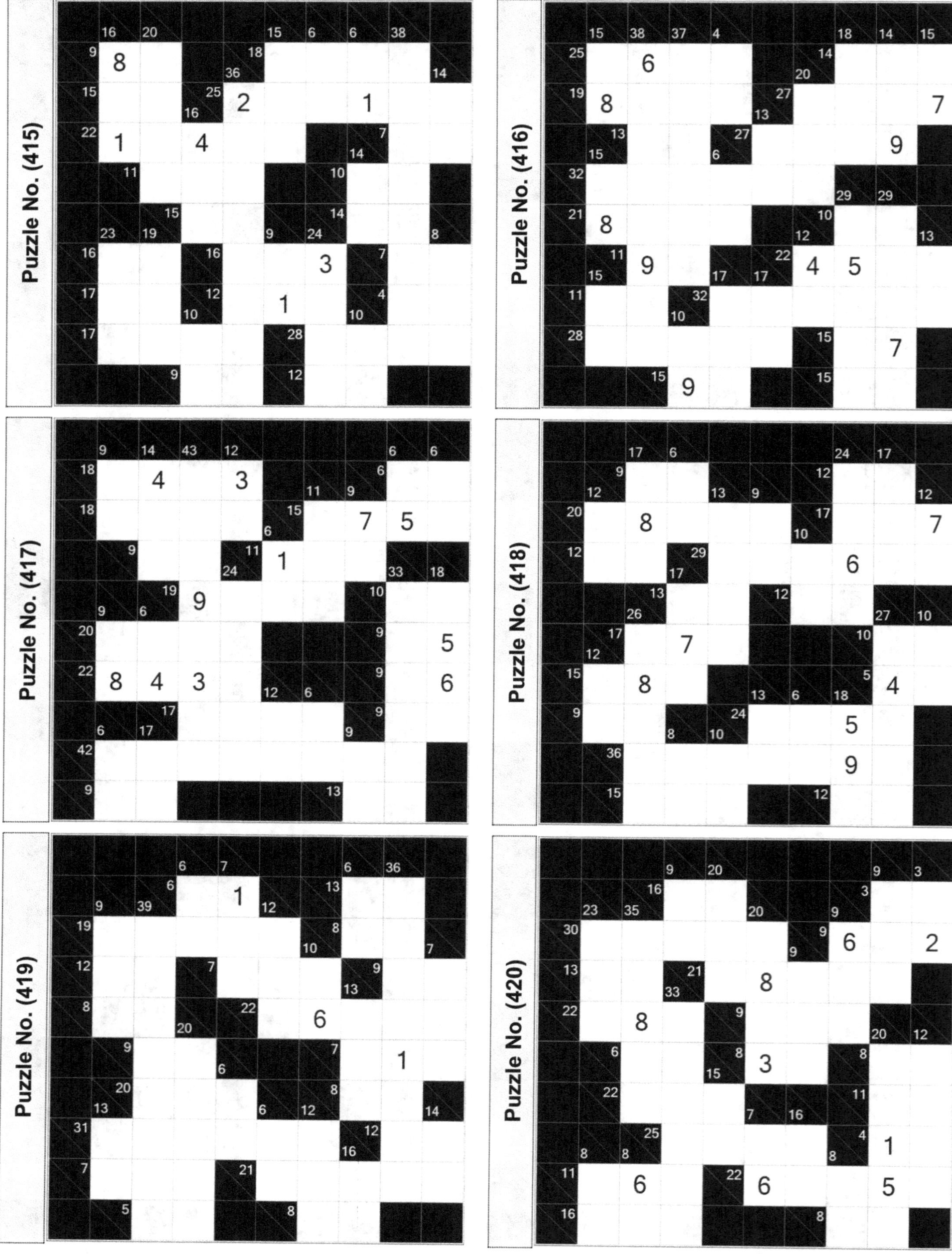

Puzzle No. (415)

Puzzle No. (416)

Puzzle No. (417)

Puzzle No. (418)

Puzzle No. (419)

Puzzle No. (420)

Solution on Page (190)

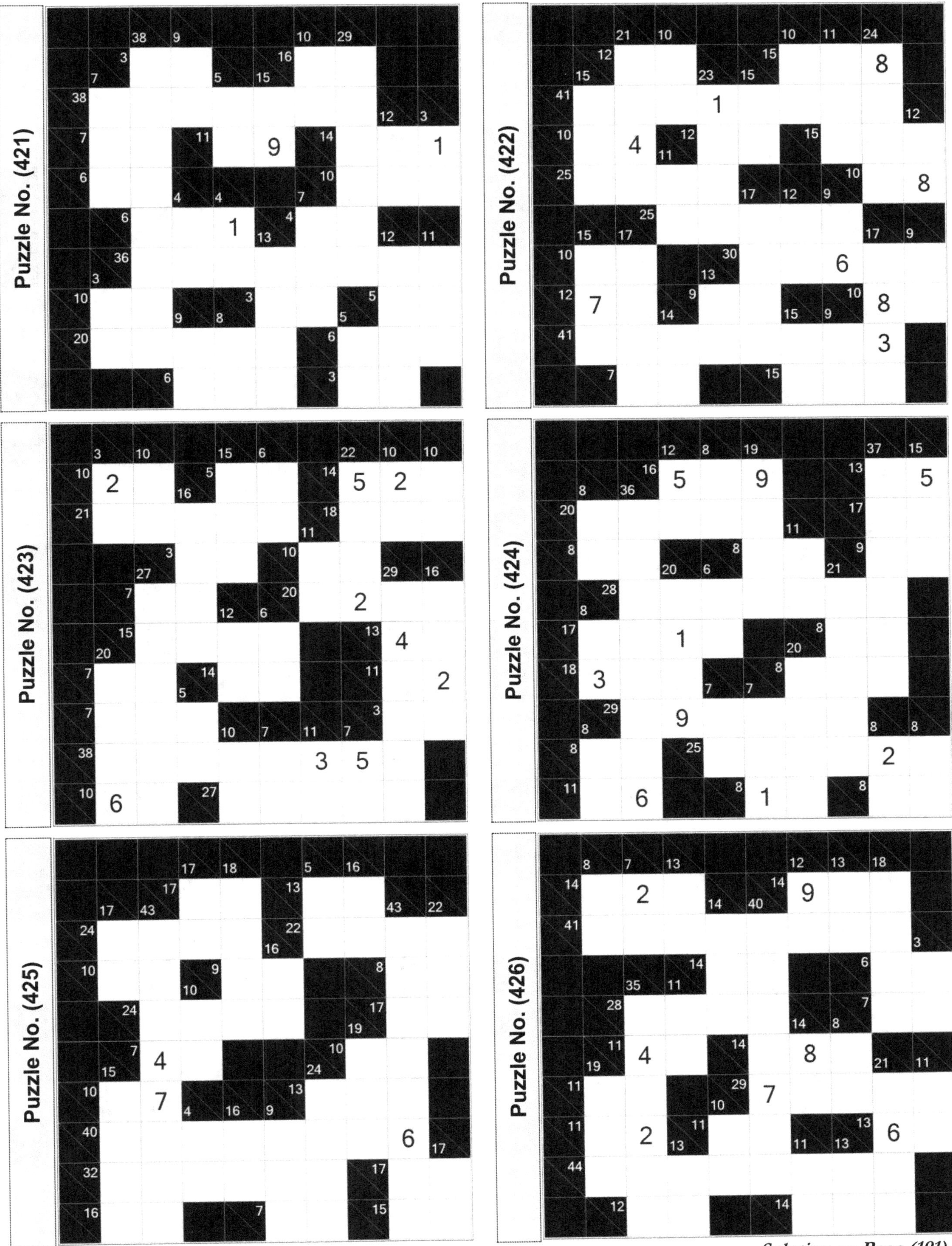

Solution on Page (191)

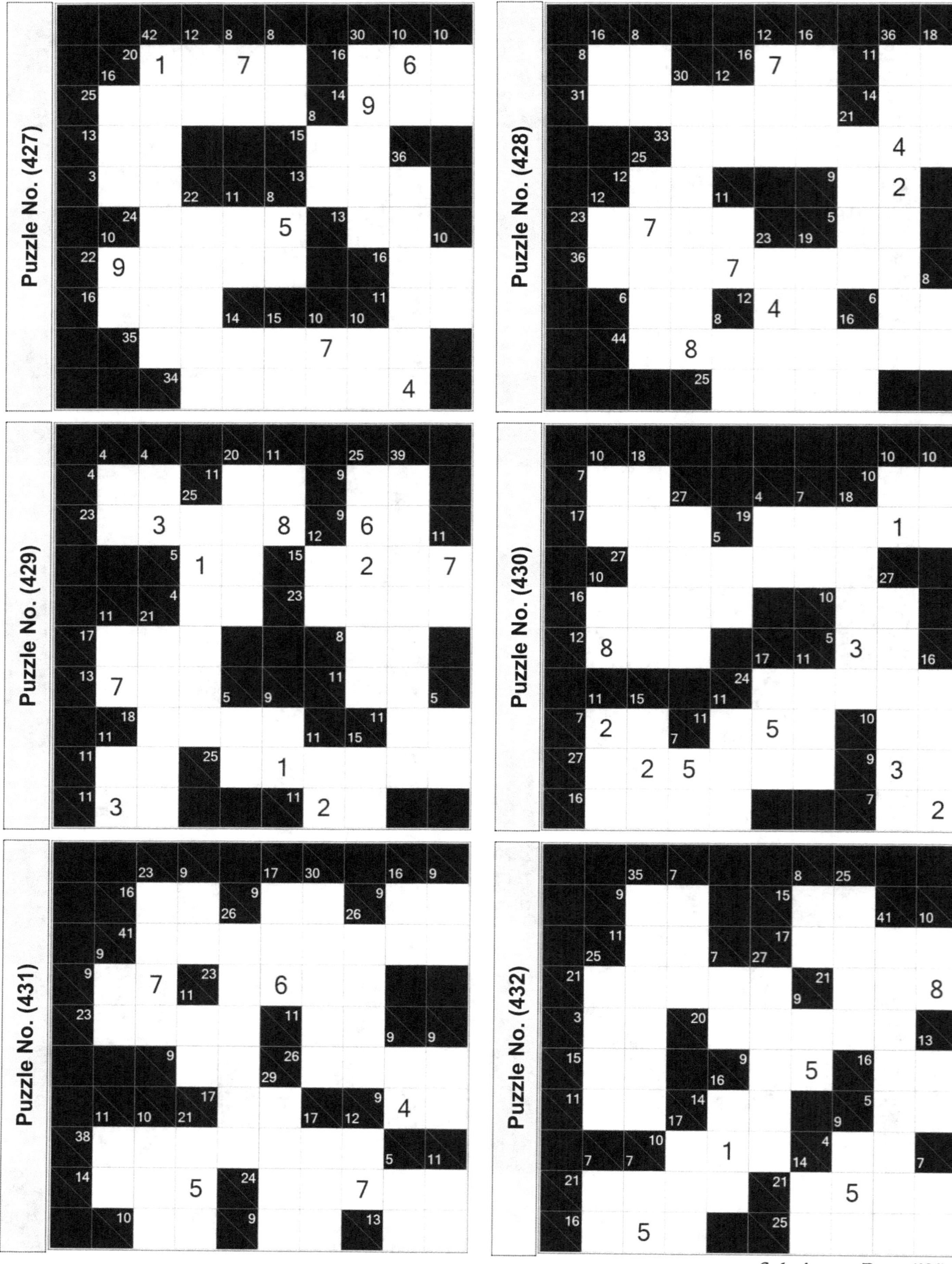

Solution on Page (191)

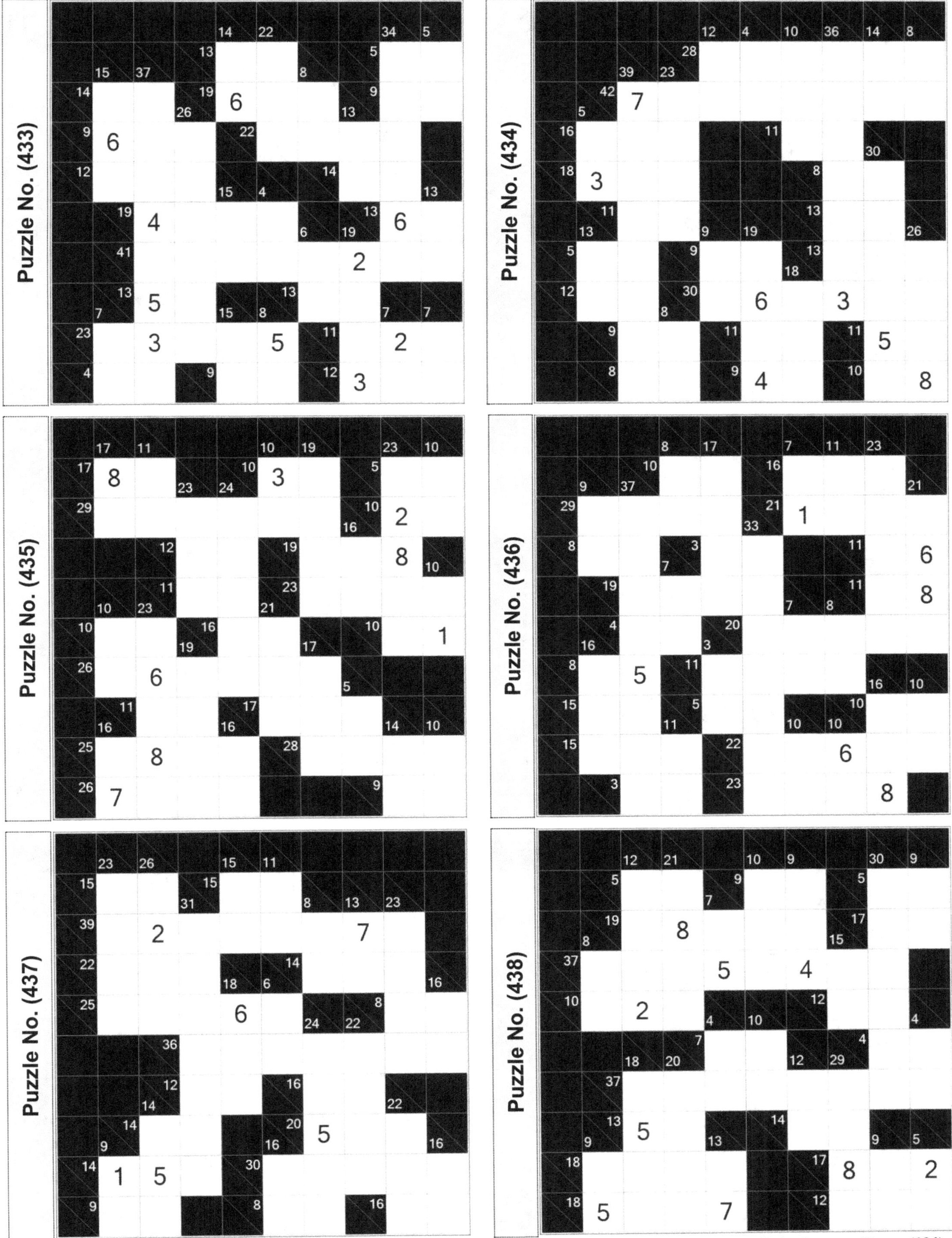

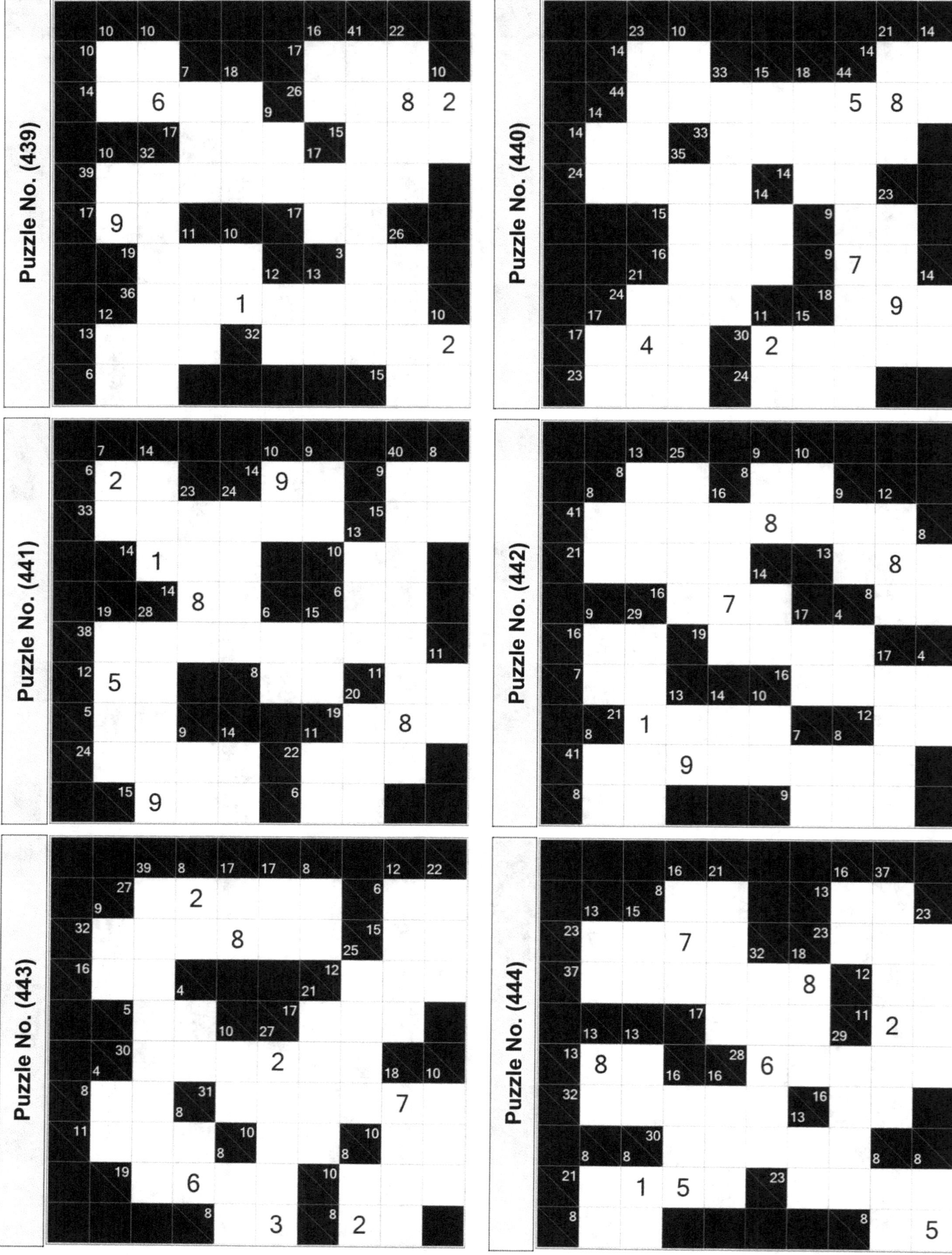

Puzzle No. (439)
Puzzle No. (440)
Puzzle No. (441)
Puzzle No. (442)
Puzzle No. (443)
Puzzle No. (444)
Solution on Pages (191-192)

(77)

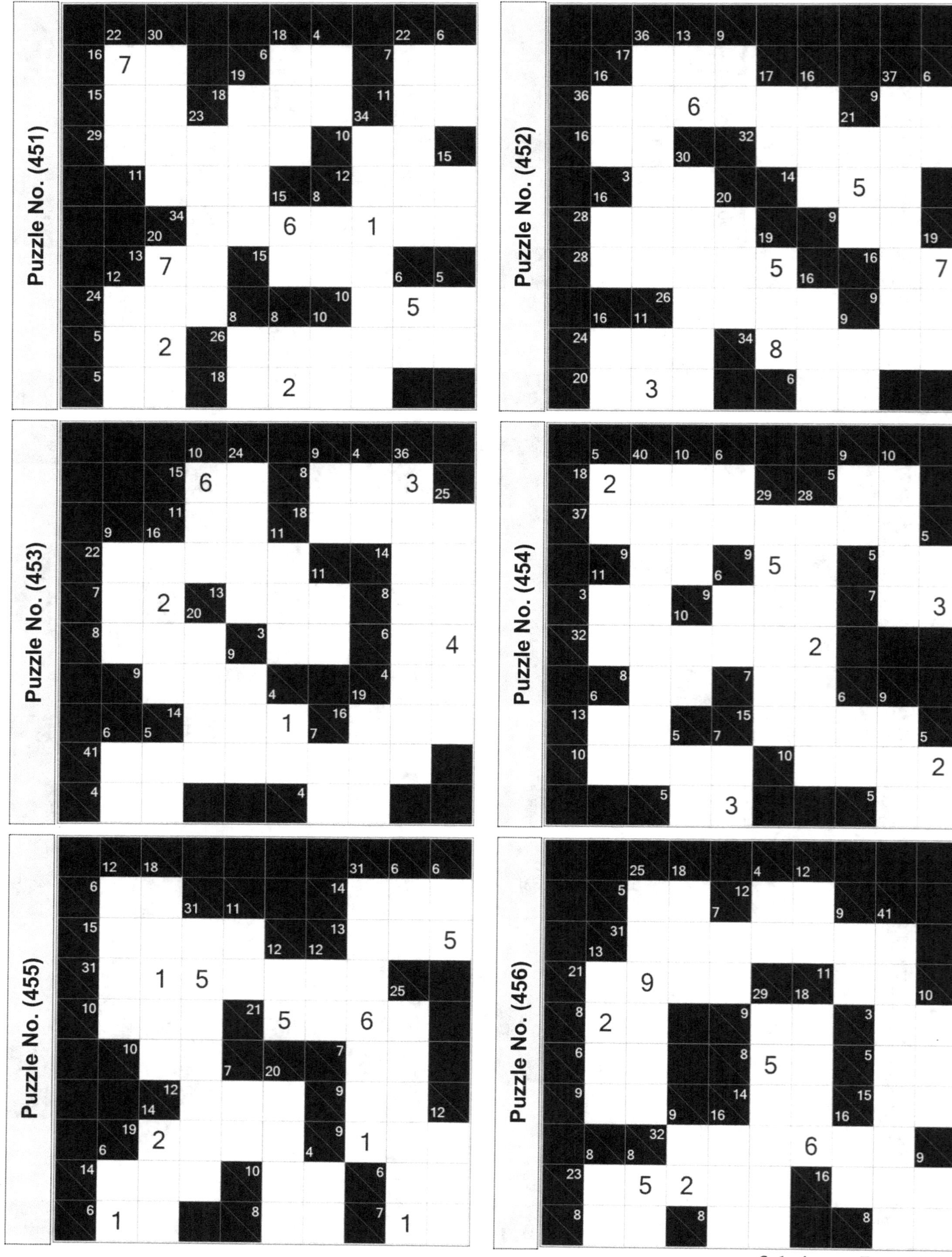

Solution on Page (192)

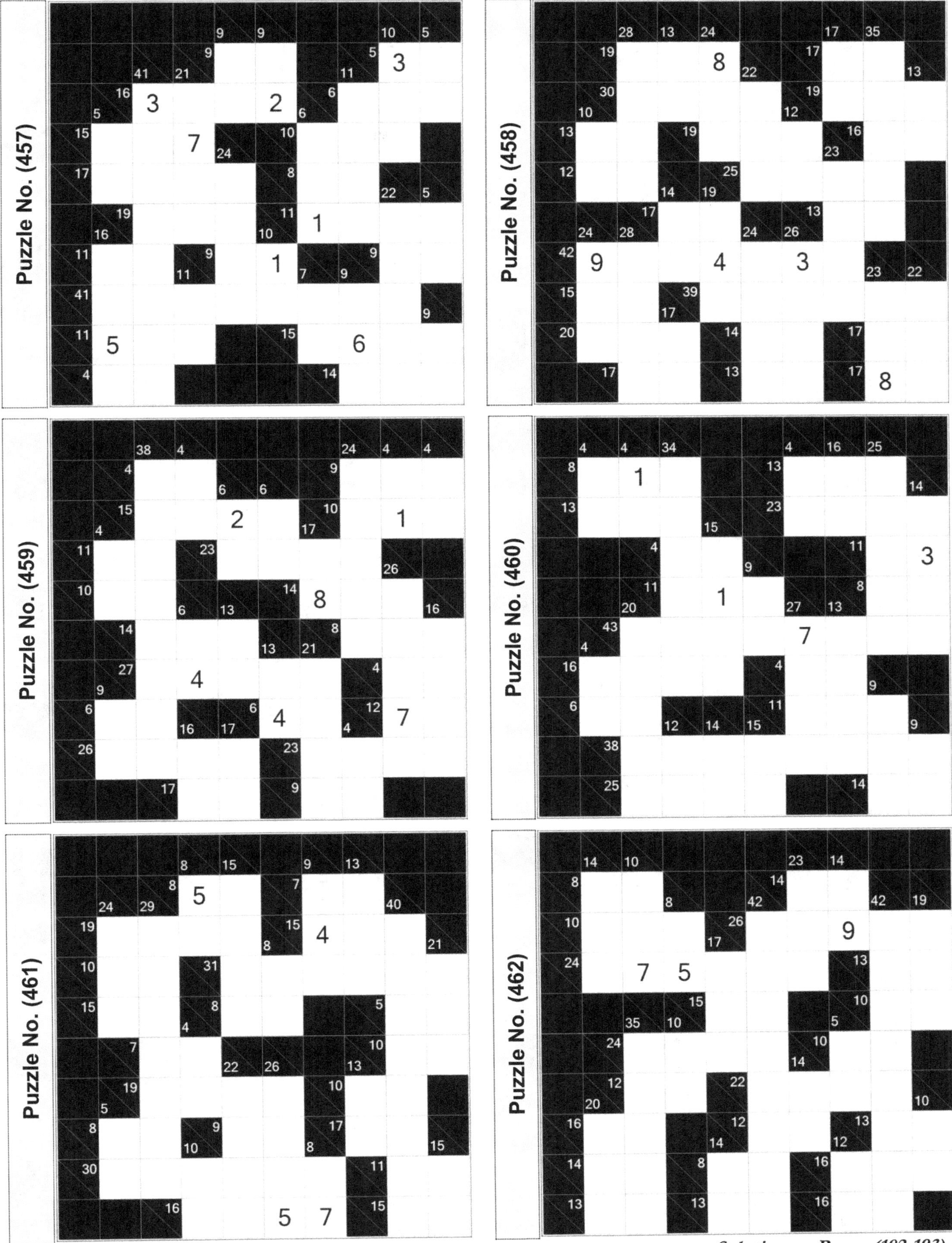

Puzzle No. (457)
Puzzle No. (458)
Puzzle No. (459)
Puzzle No. (460)
Puzzle No. (461)
Puzzle No. (462)
Solution on Pages (192-193)

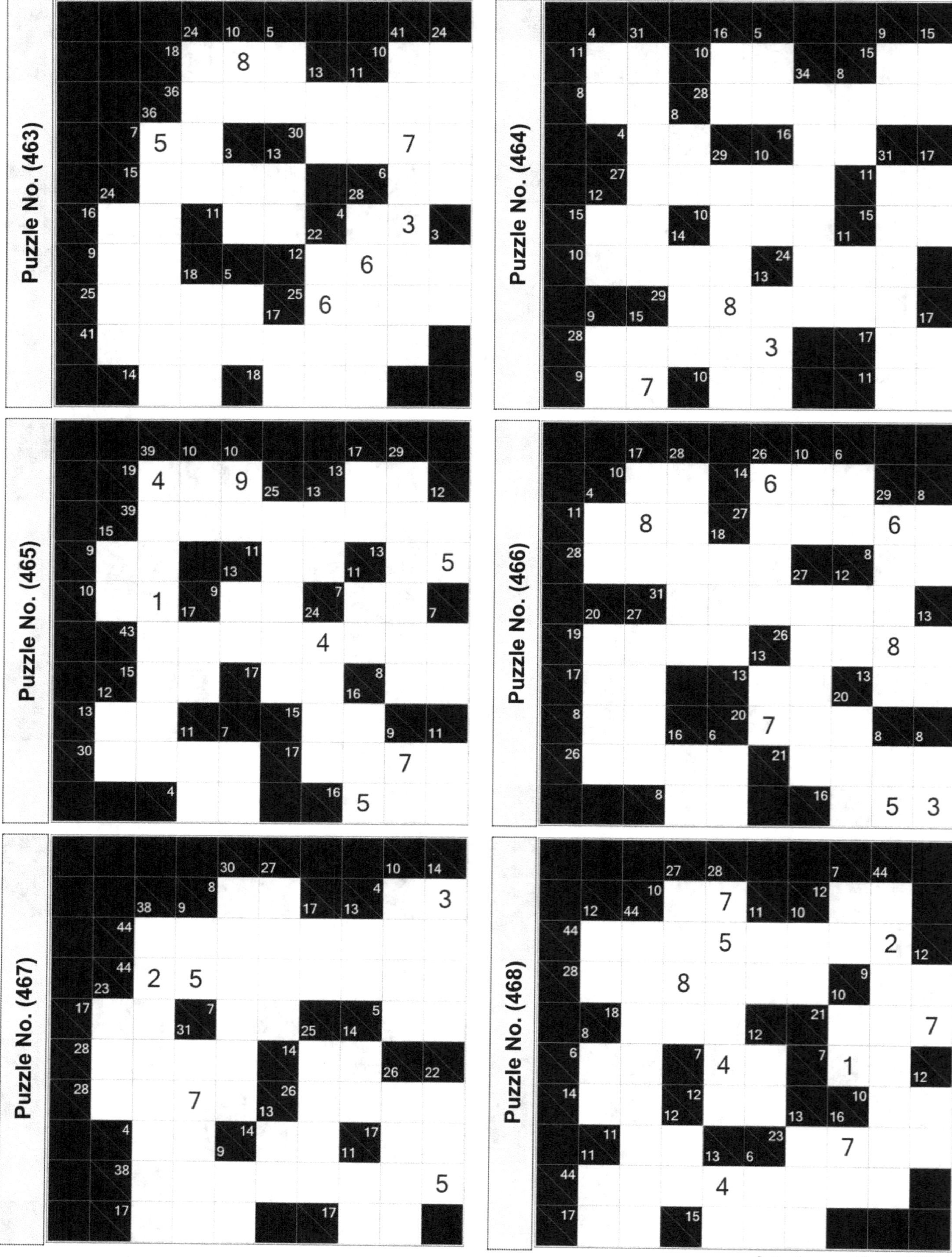

Solution on Page (193)

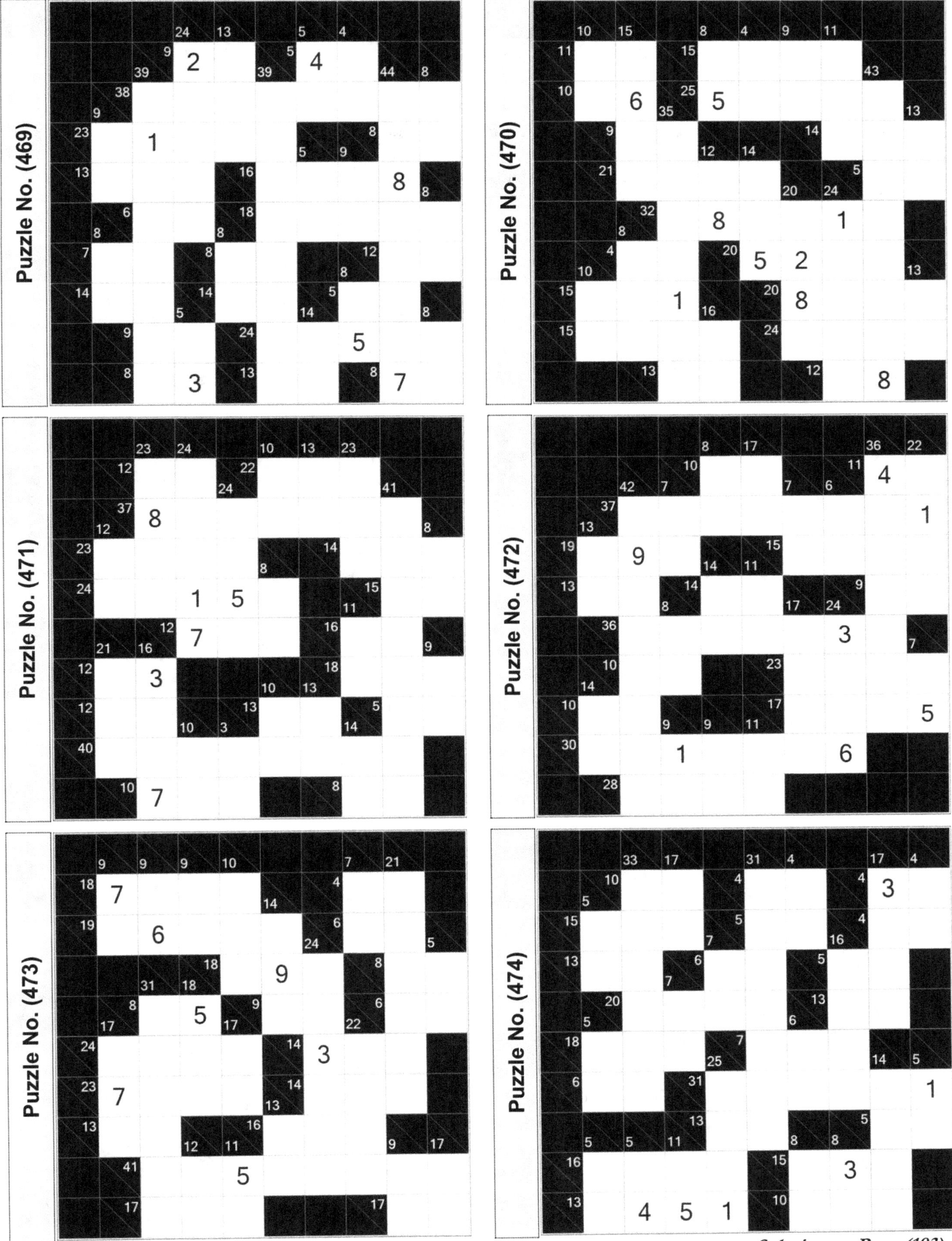

Puzzle No. (469)

Puzzle No. (470)

Puzzle No. (471)

Puzzle No. (472)

Puzzle No. (473)

Puzzle No. (474)

Solution on Page (193)

(81)

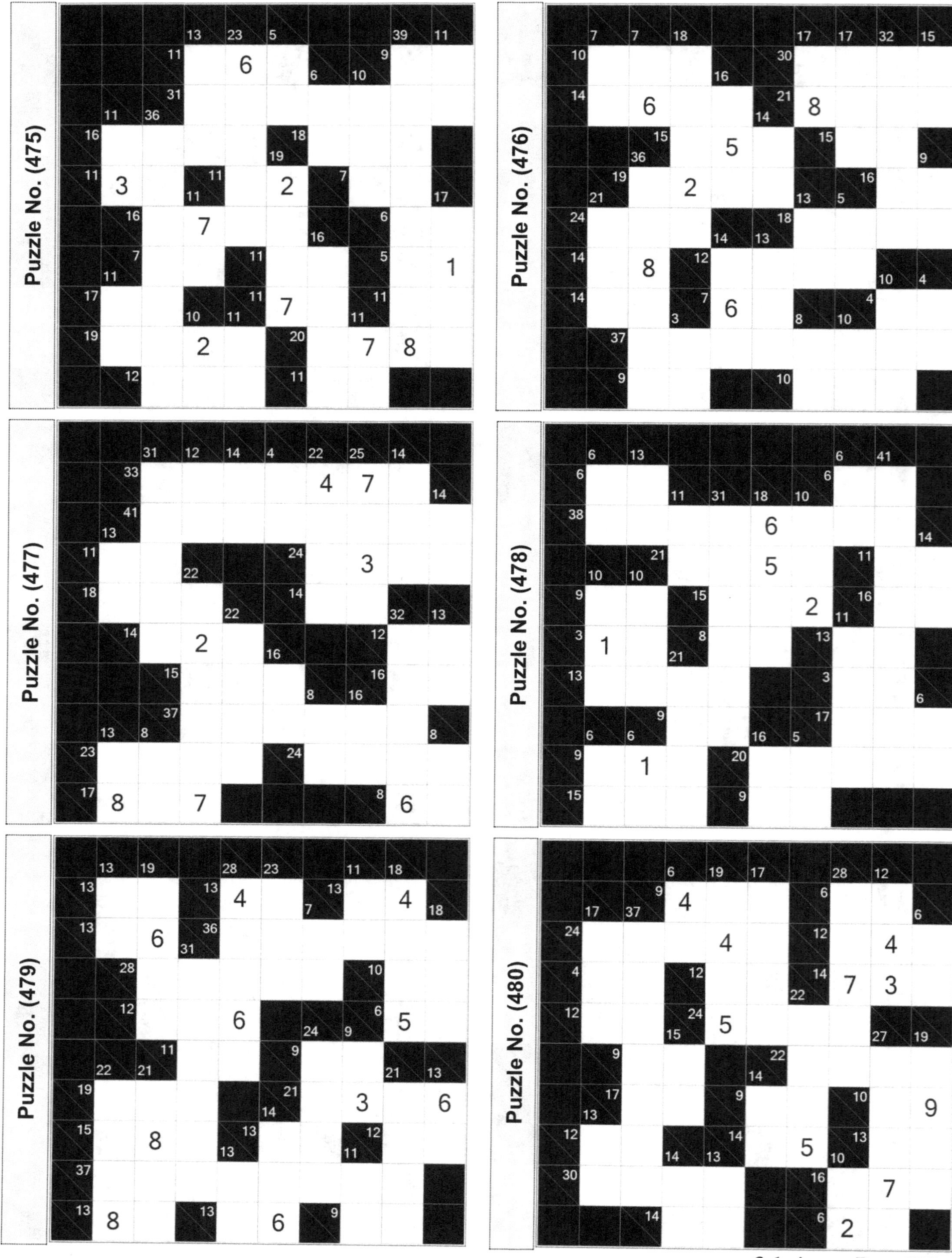

Puzzle No. (475)
Puzzle No. (476)
Puzzle No. (477)
Puzzle No. (478)
Puzzle No. (479)
Puzzle No. (480)
Solution on Page (193)

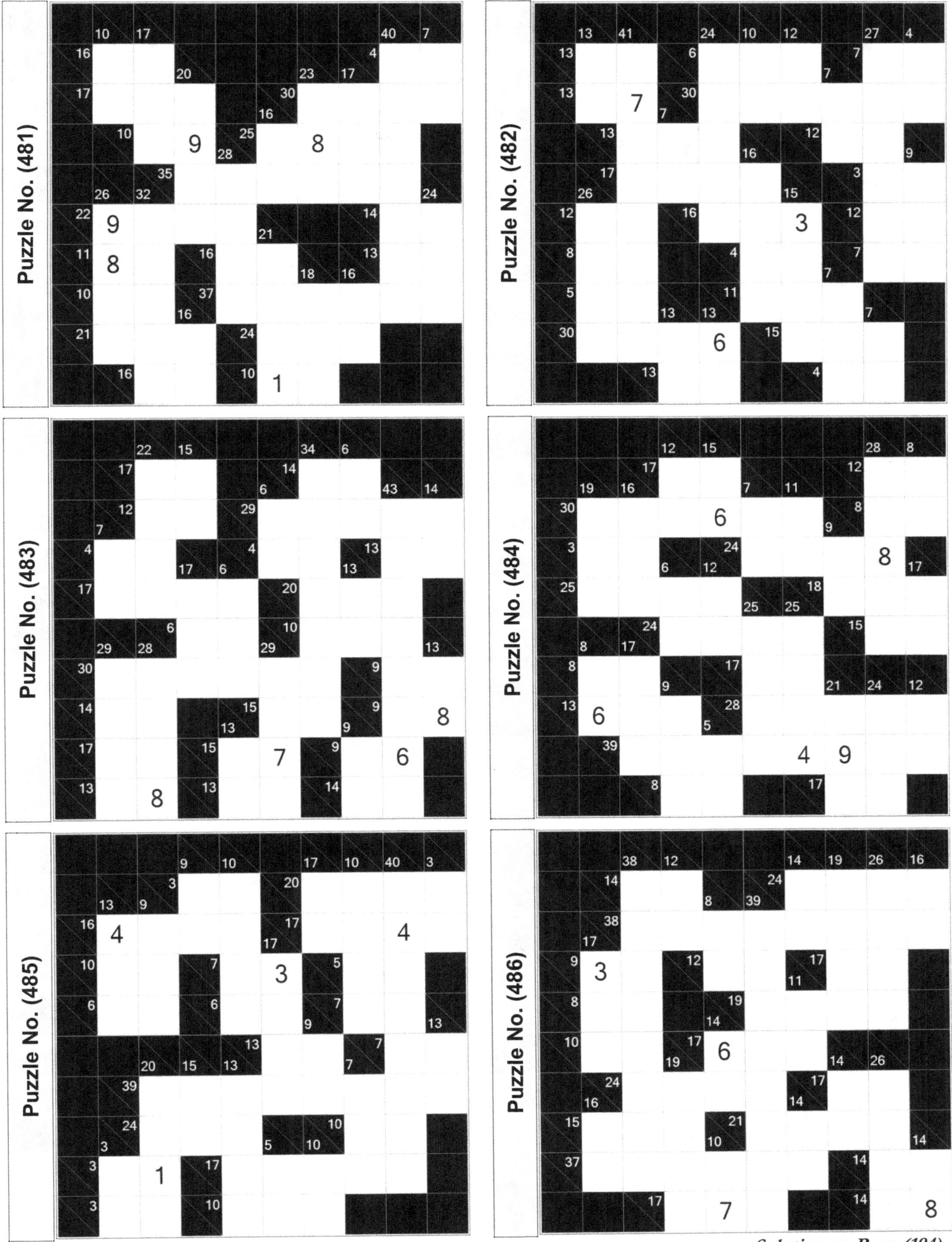

Solution on Page (194)

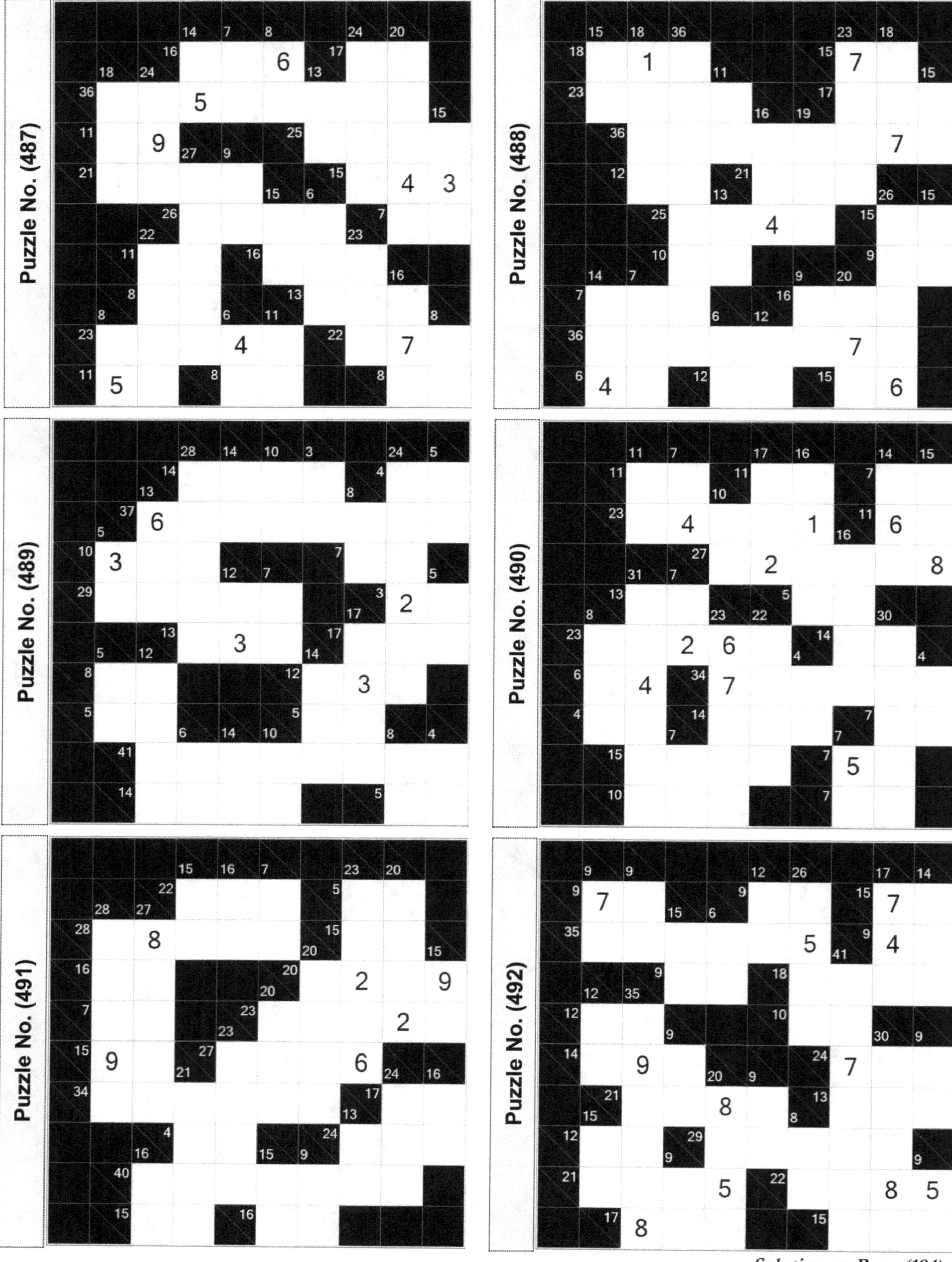

Solution on Page (194)

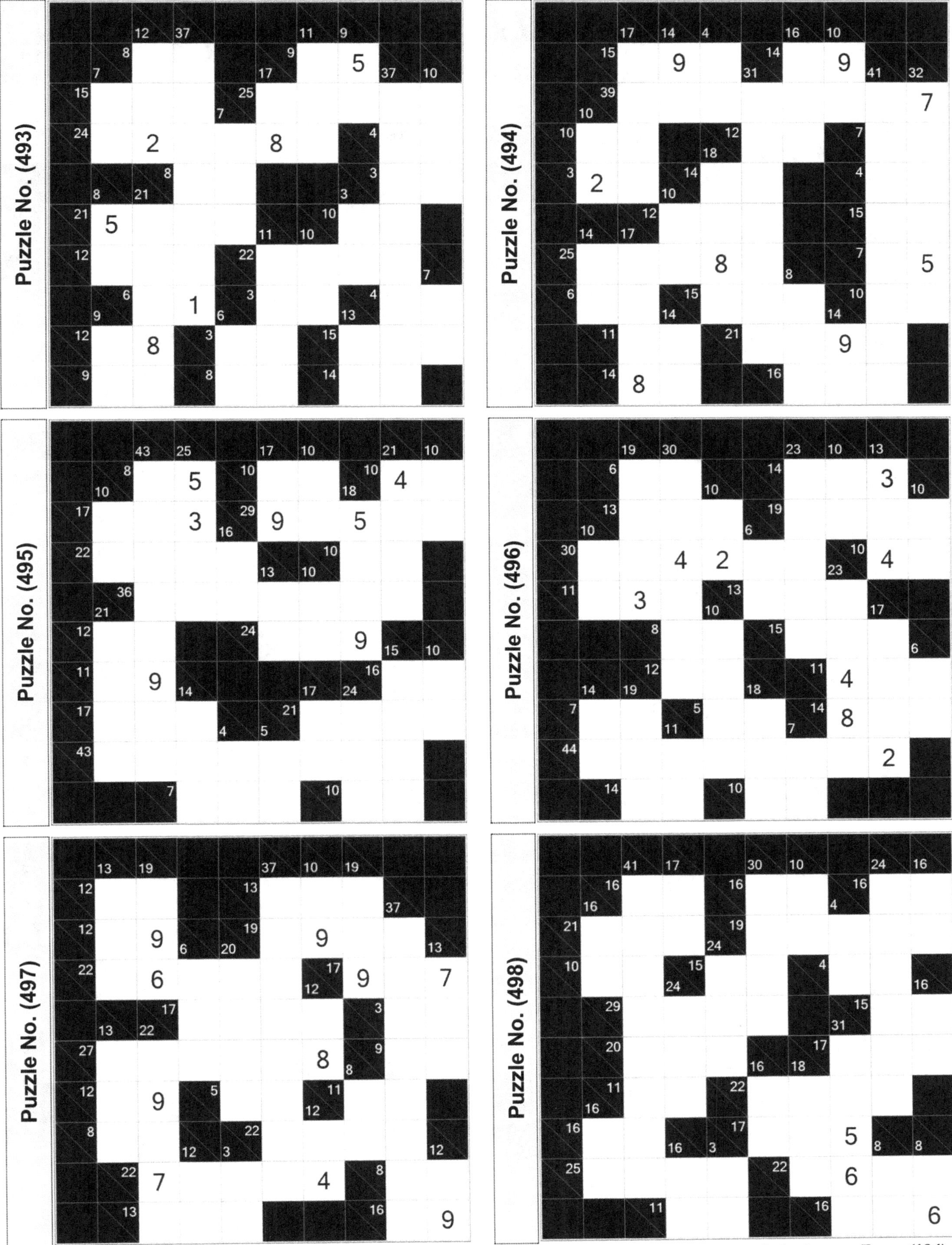

Puzzle No. (493)
Puzzle No. (494)
Puzzle No. (495)
Puzzle No. (496)
Puzzle No. (497)
Puzzle No. (498)
Solution on Page (194)

Solution on Pages (194-195)

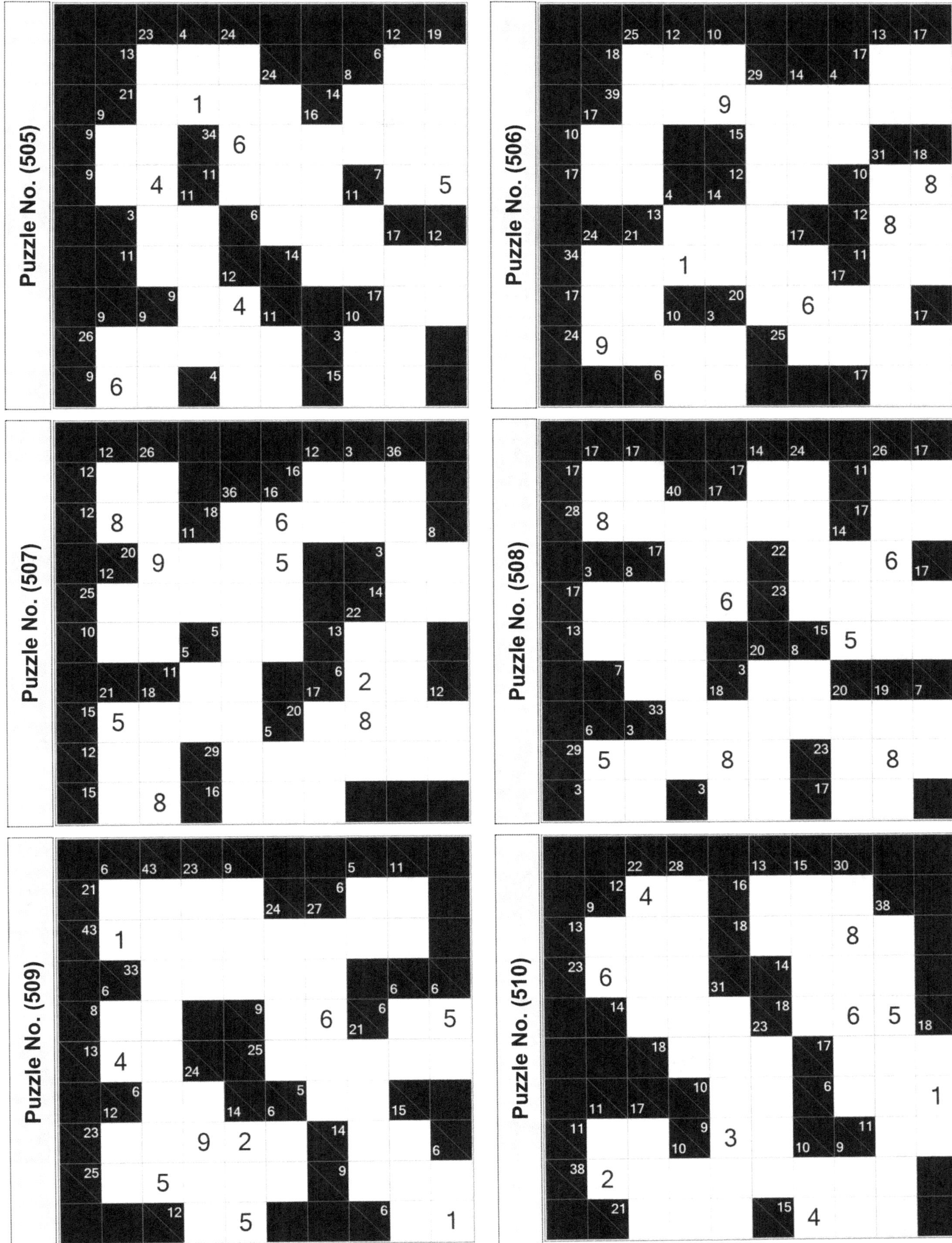

Puzzle No. (505)

Puzzle No. (506)

Puzzle No. (507)

Puzzle No. (508)

Puzzle No. (509)

Puzzle No. (510)

Solution on Page (195)

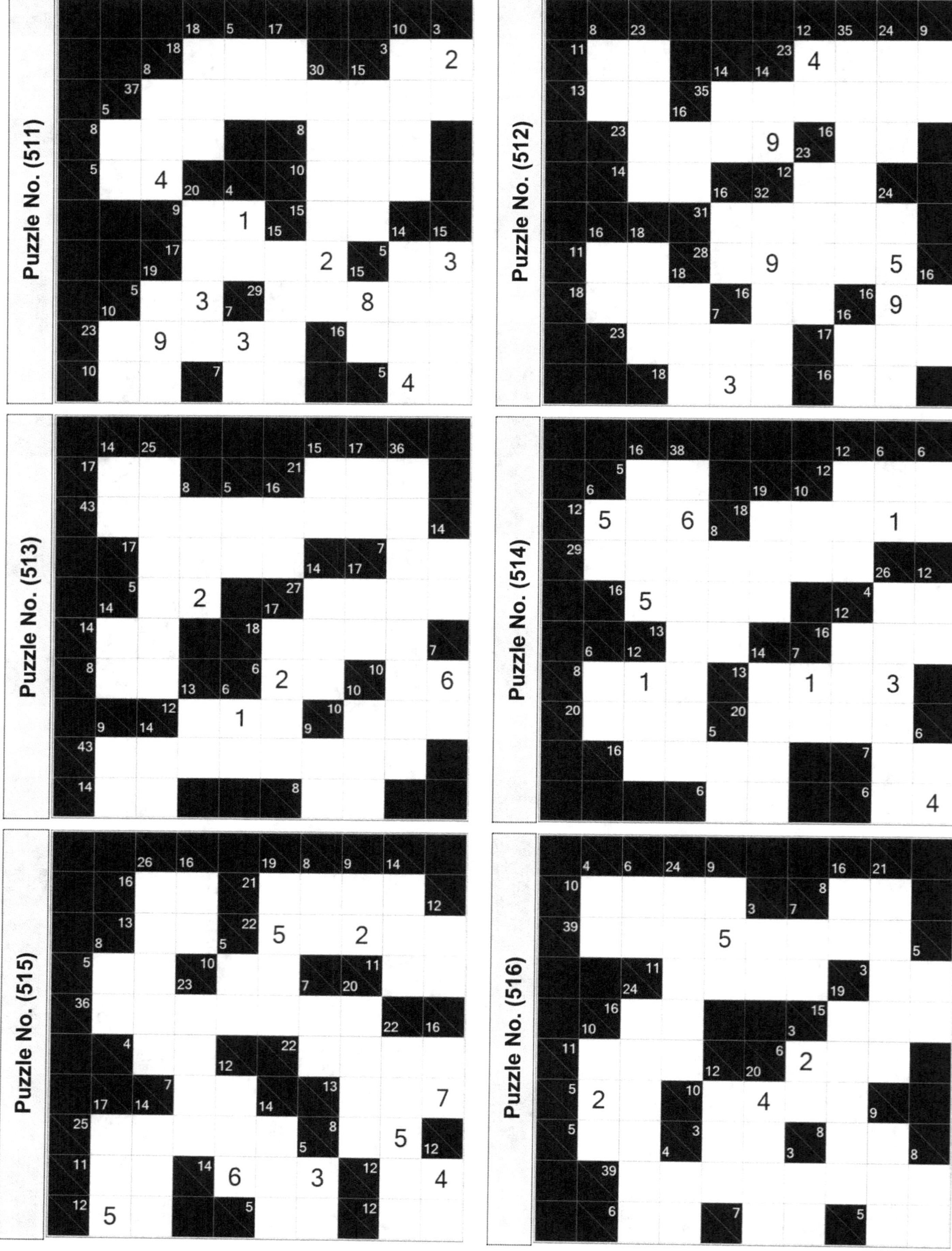

Solution on Page (195)

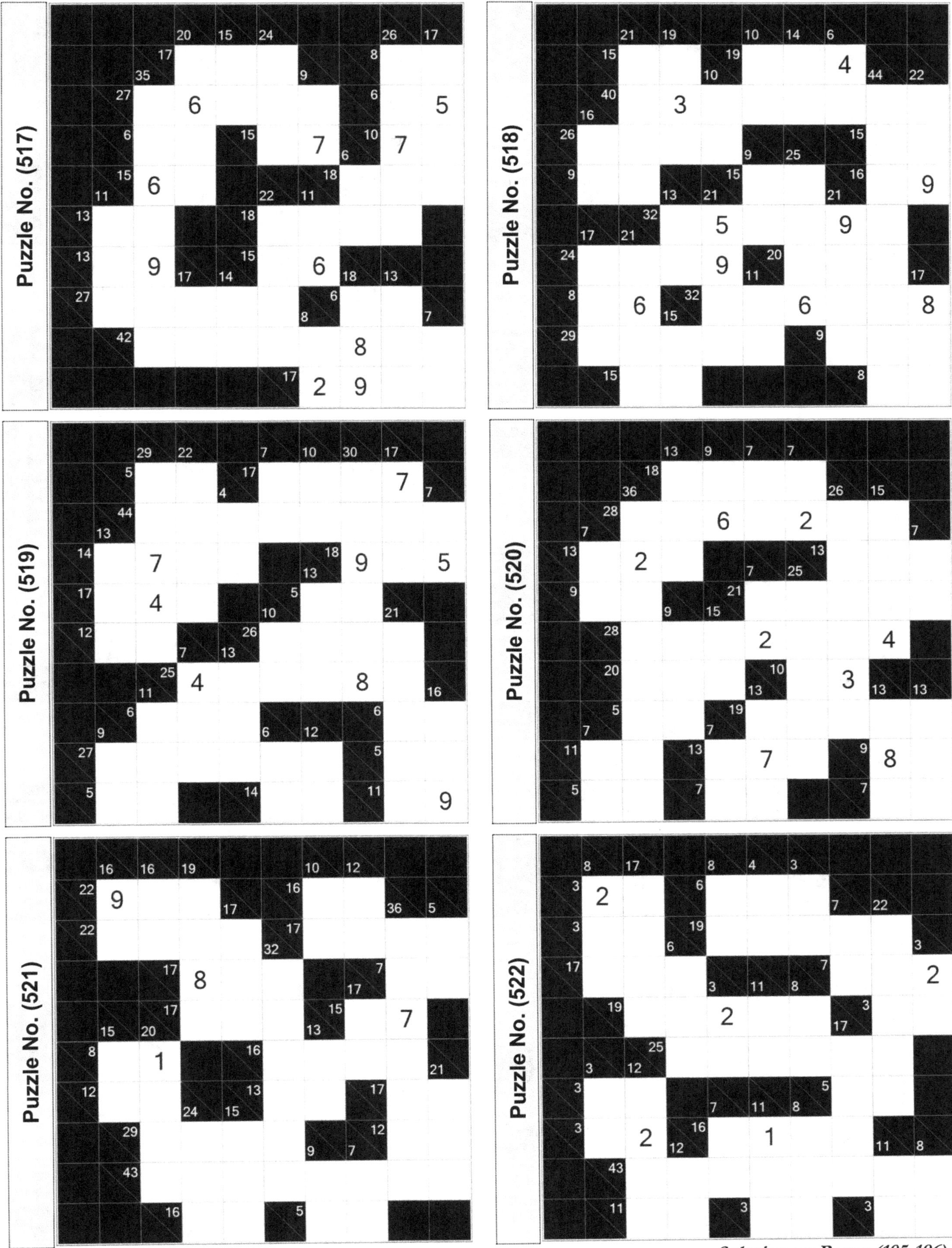

Solution on Pages (195-196)

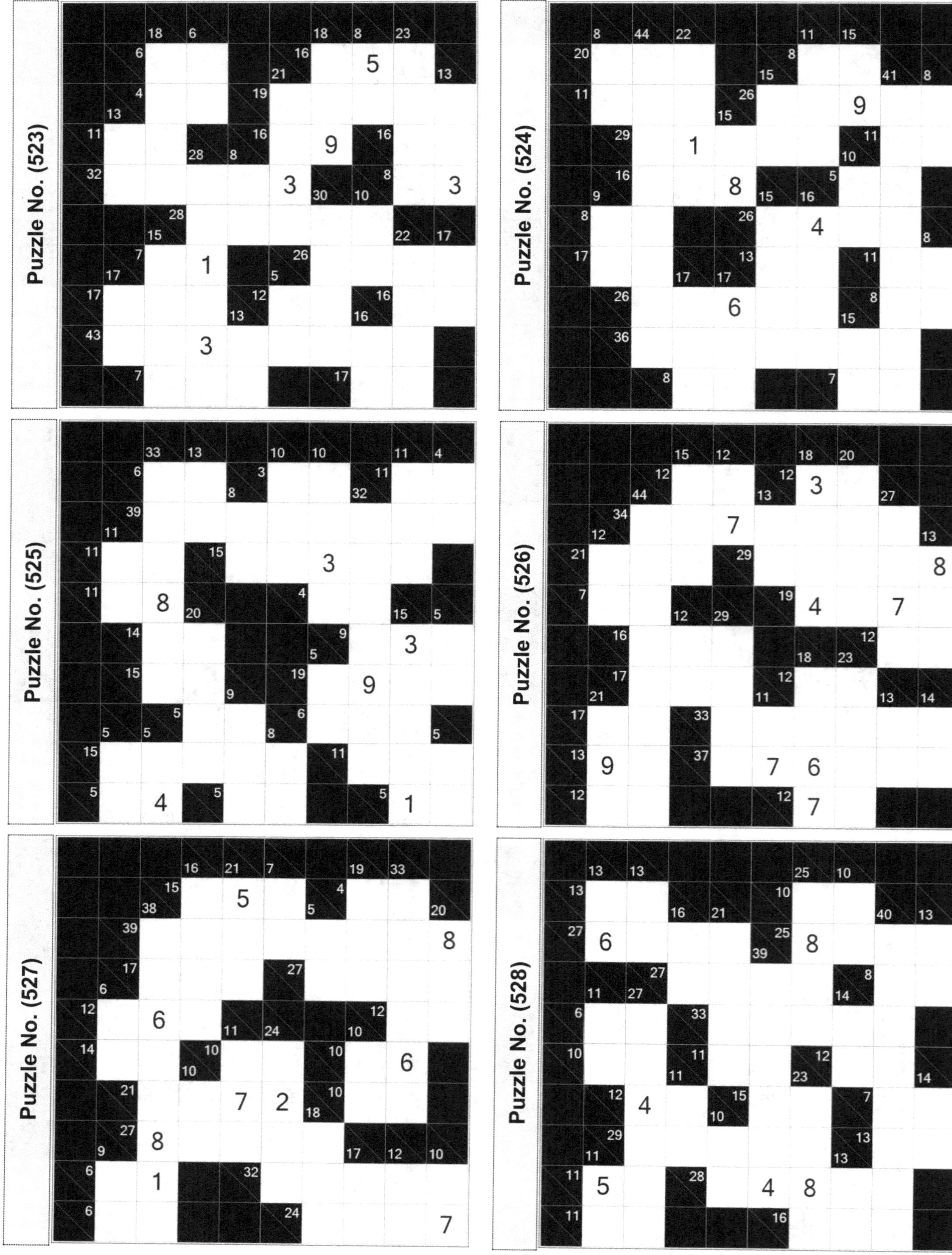

Puzzle No. (523)

Puzzle No. (524)

Puzzle No. (525)

Puzzle No. (526)

Puzzle No. (527)

Puzzle No. (528)

Solution on Page (196)

(90)

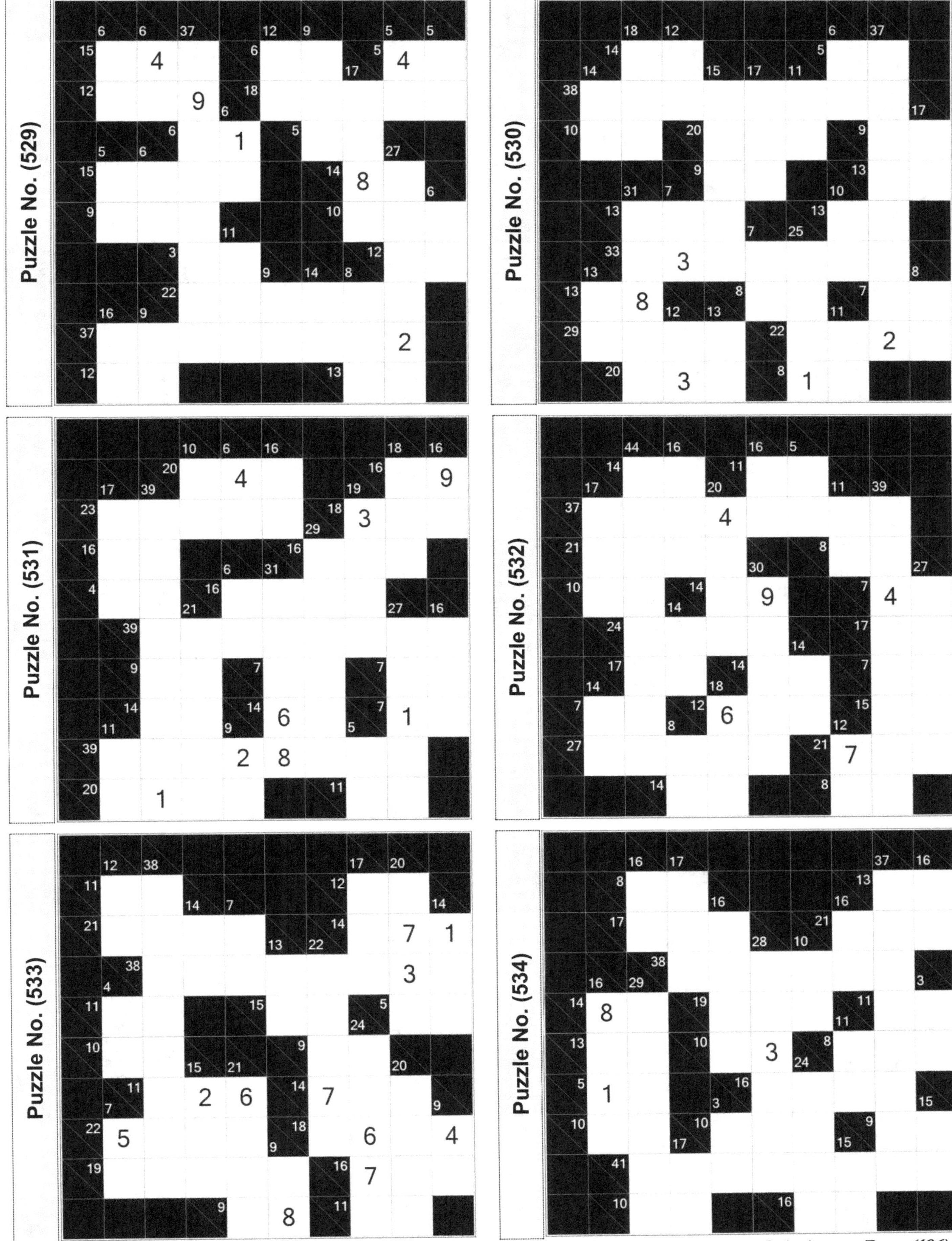

Puzzle No. (529)
Puzzle No. (530)
Puzzle No. (531)
Puzzle No. (532)
Puzzle No. (533)
Puzzle No. (534)

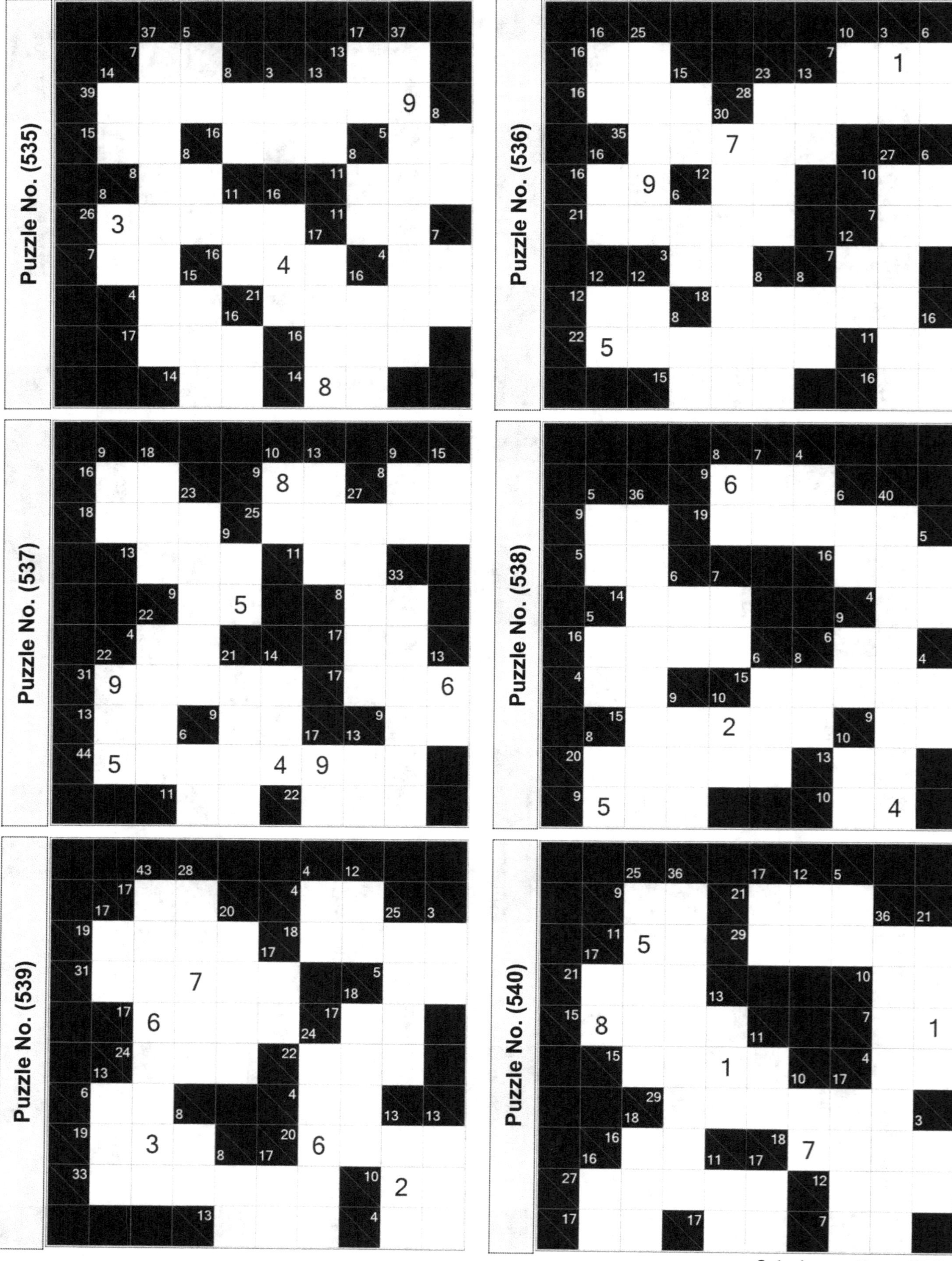

Puzzle No. (535)

Puzzle No. (536)

Puzzle No. (537)

Puzzle No. (538)

Puzzle No. (539)

Puzzle No. (540)

Solution on Page (196)

Solution on Page (197)

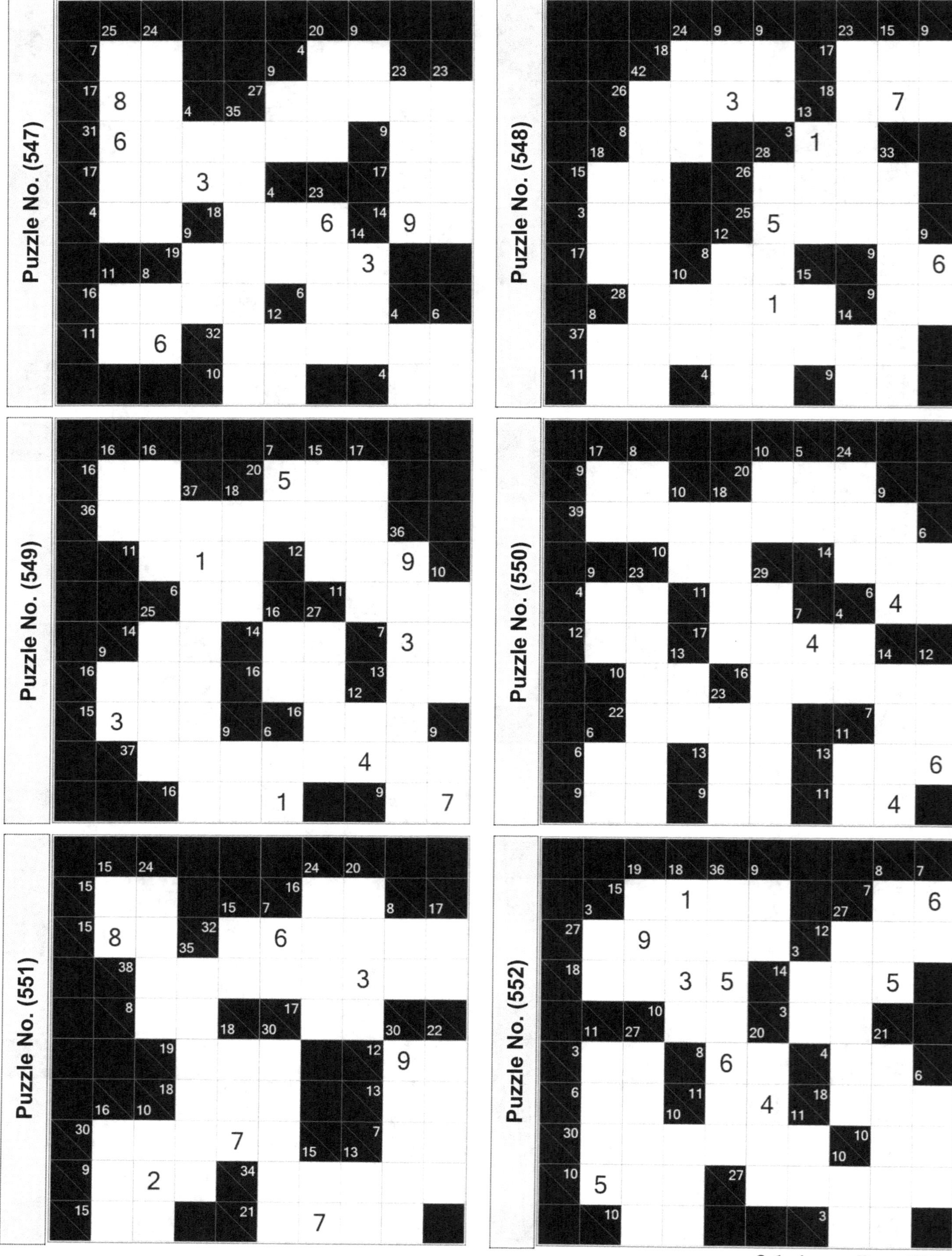

Puzzle No. (547)

Puzzle No. (548)

Puzzle No. (549)

Puzzle No. (550)

Puzzle No. (551)

Puzzle No. (552)

Solution on Page (197)

Solution on Page (197)

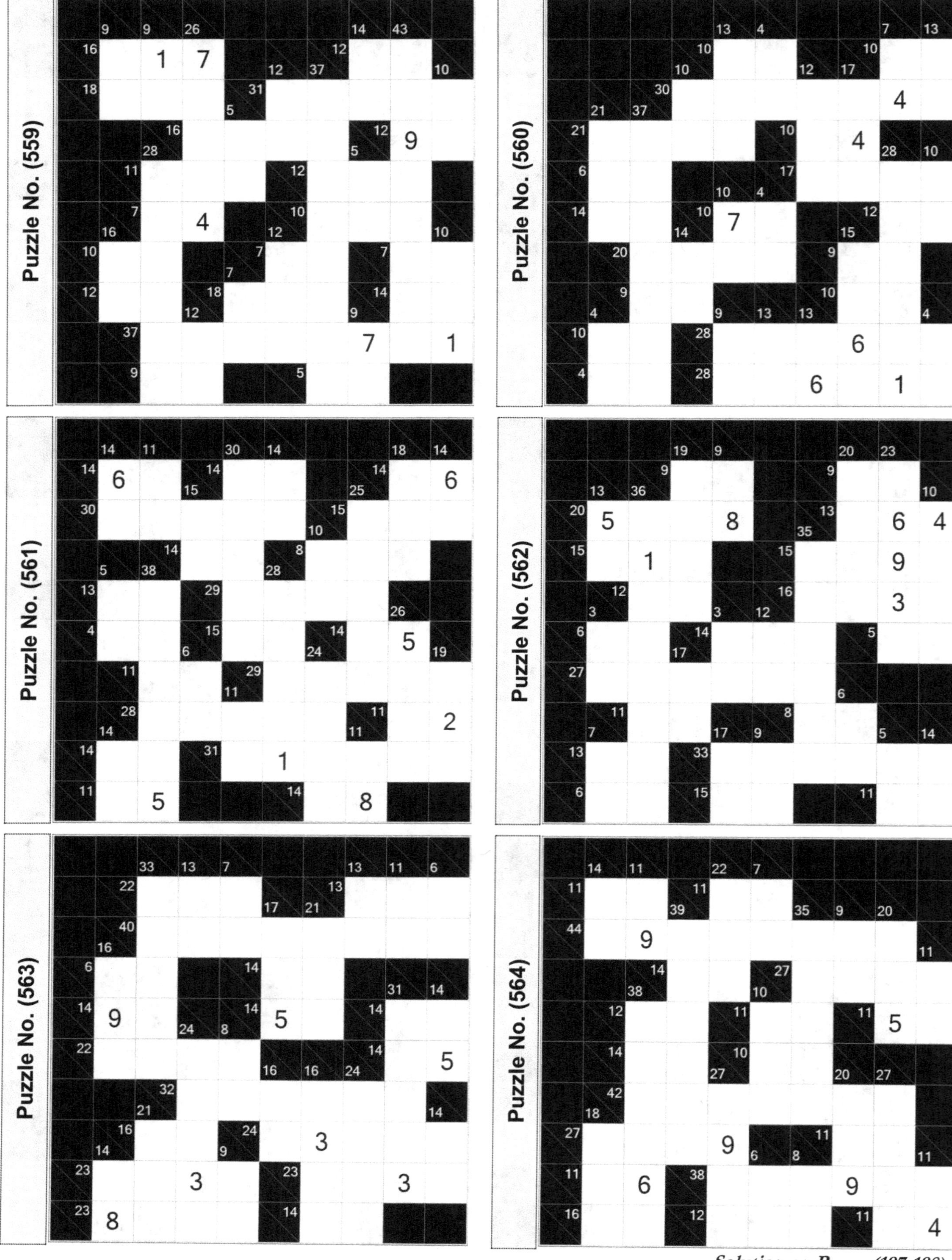

Solution on Pages (197-198)

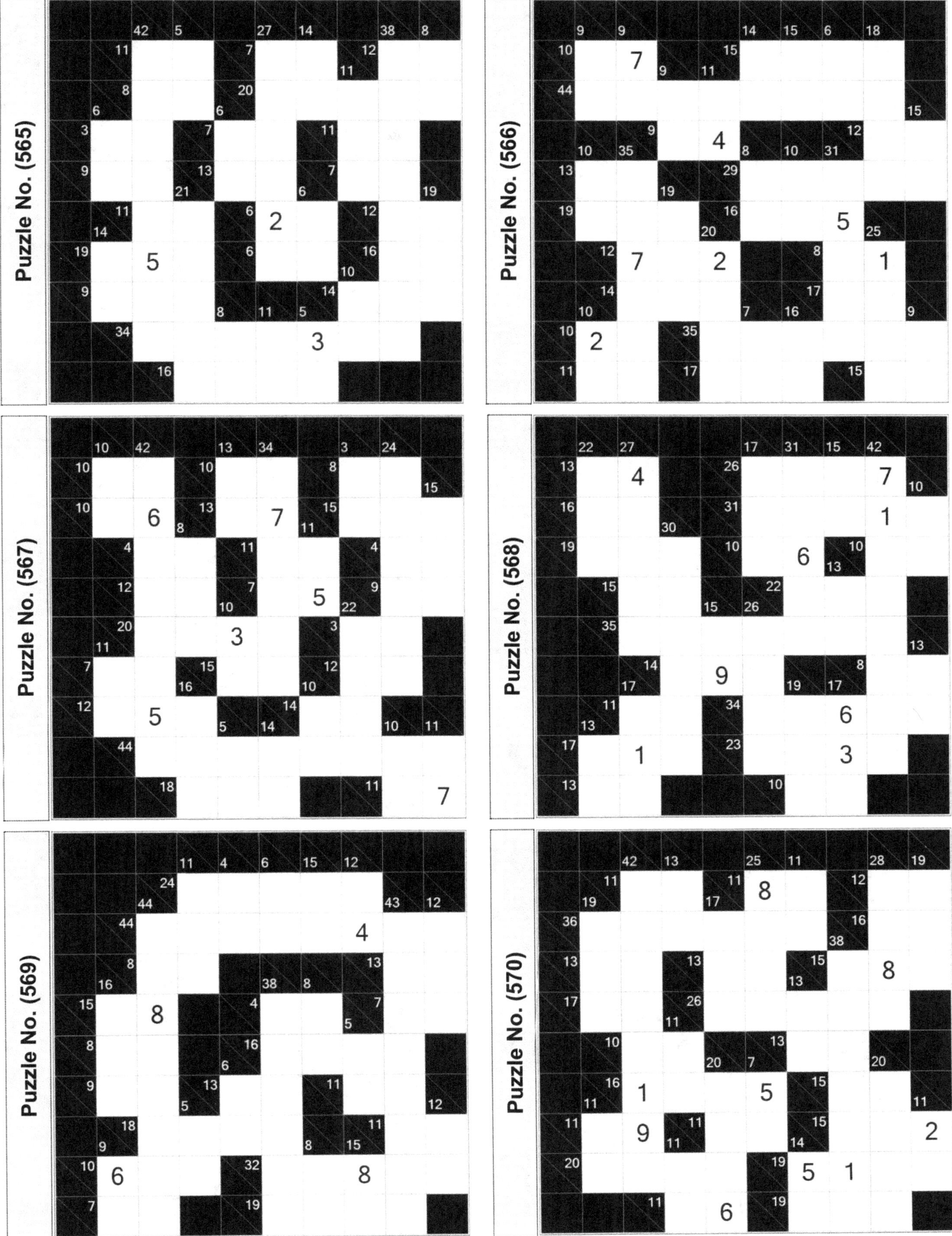

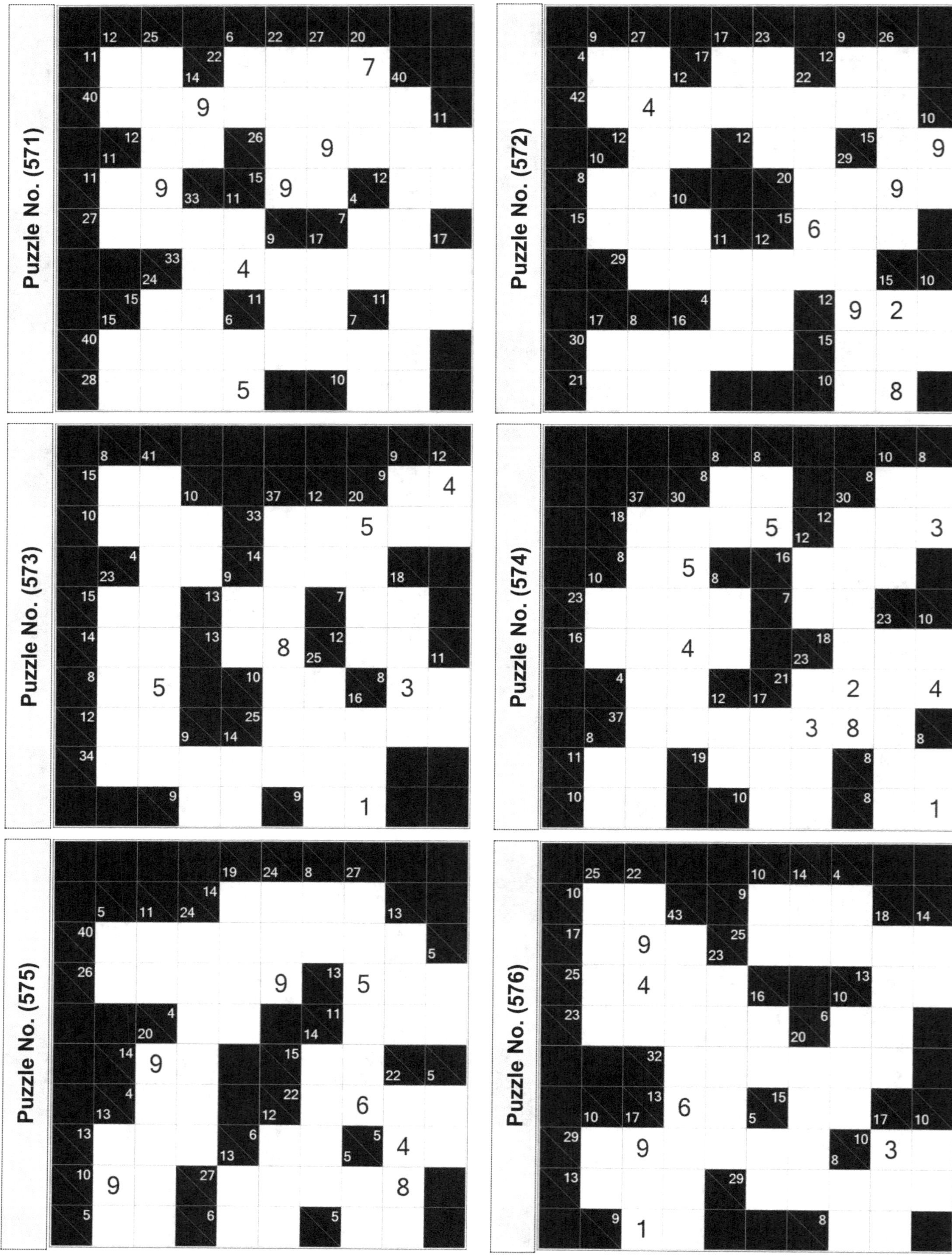

Puzzle No. (571) **Puzzle No. (572)**

Puzzle No. (573) **Puzzle No. (574)**

Puzzle No. (575) **Puzzle No. (576)**

Solution on Page (198)

Puzzle No. (577)
Puzzle No. (578)
Puzzle No. (579)
Puzzle No. (580)
Puzzle No. (581)
Puzzle No. (582)
Solution on Pages (198-199)

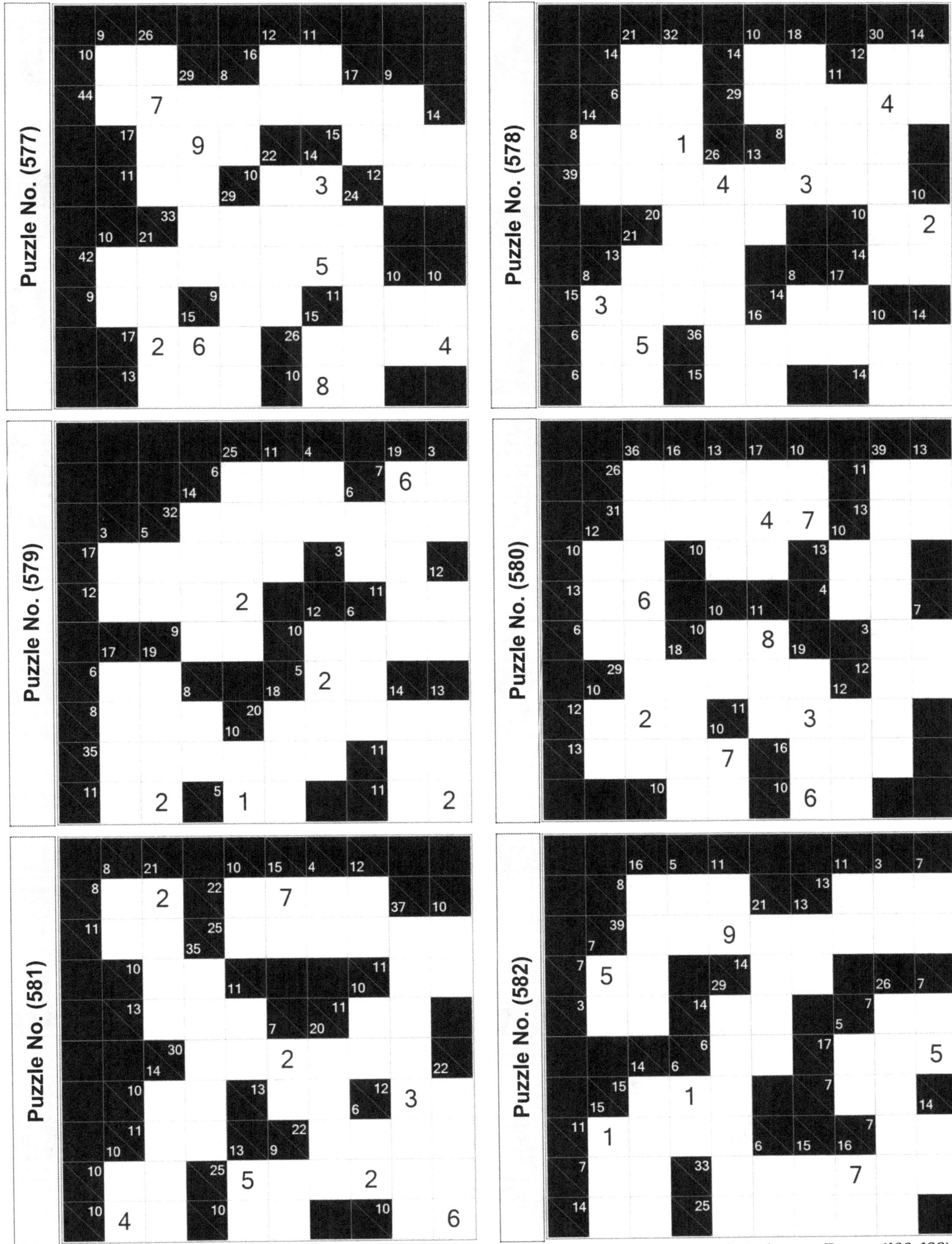

(99)

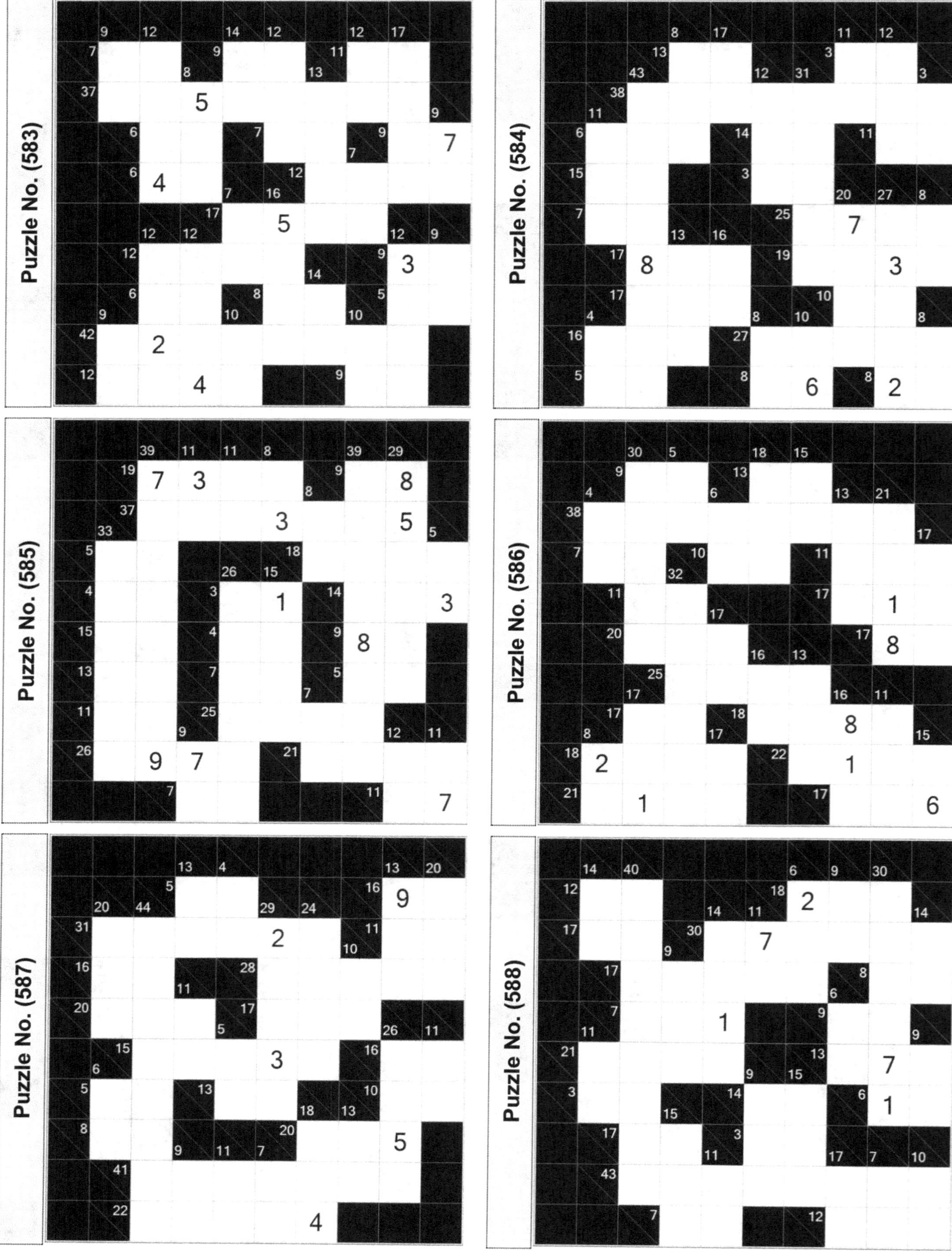

Solution on Page (199)

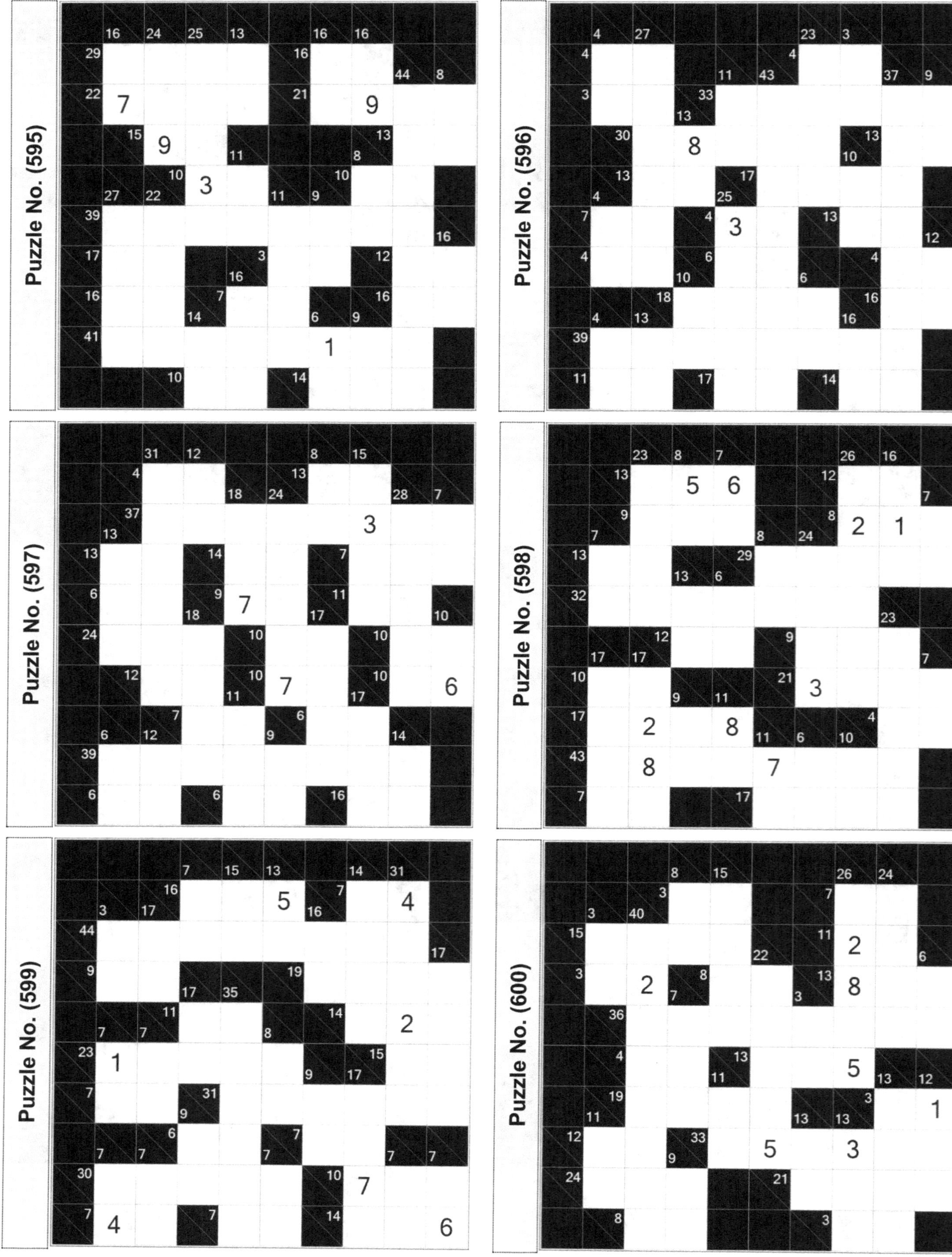

Puzzle No. (595)
Puzzle No. (596)
Puzzle No. (597)
Puzzle No. (598)
Puzzle No. (599)
Puzzle No. (600)
Solution on Page (199)

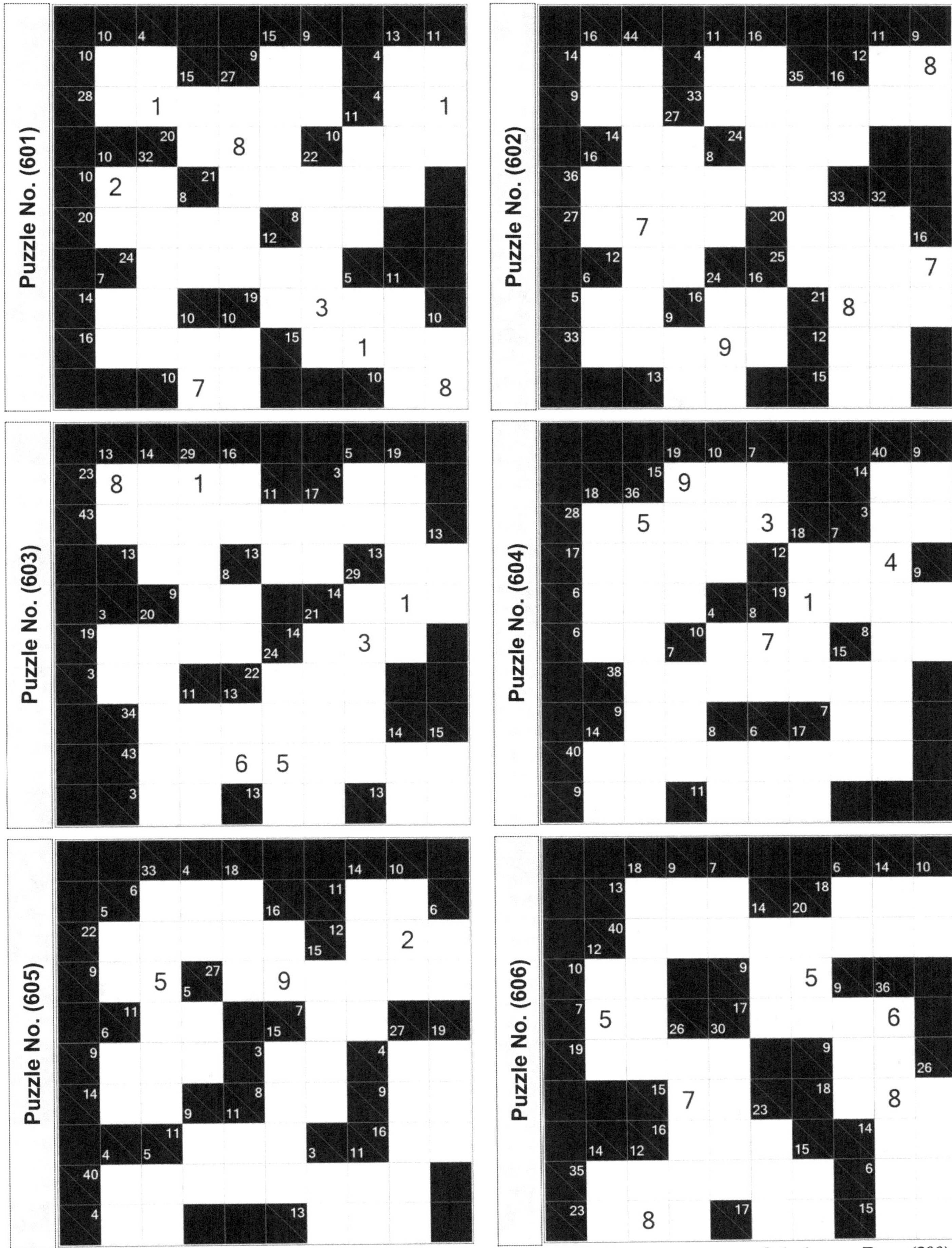

Puzzle No. (601)
Puzzle No. (602)
Puzzle No. (603)
Puzzle No. (604)
Puzzle No. (605)
Puzzle No. (606)

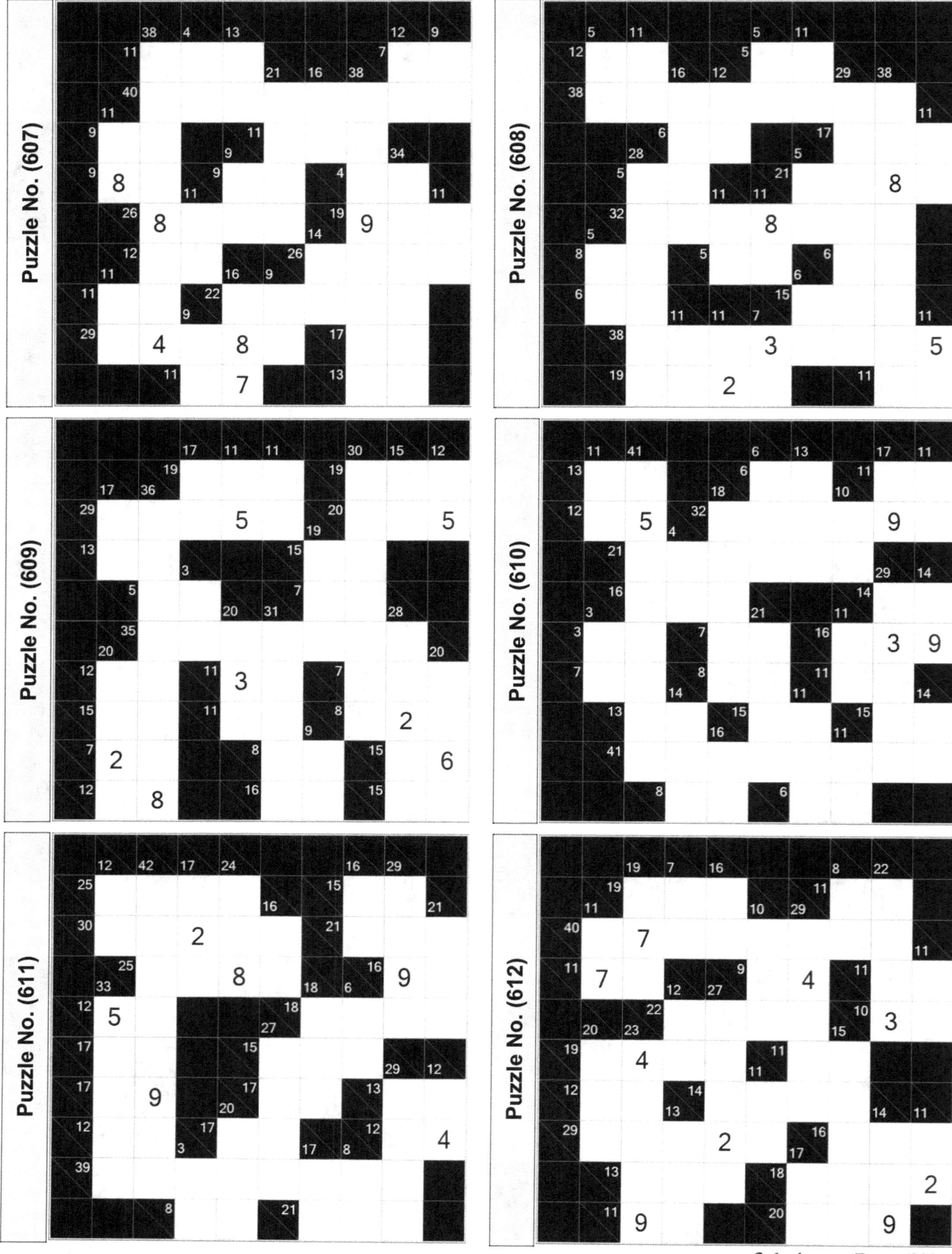

Solution on Page (200)

Solution on Page (200)

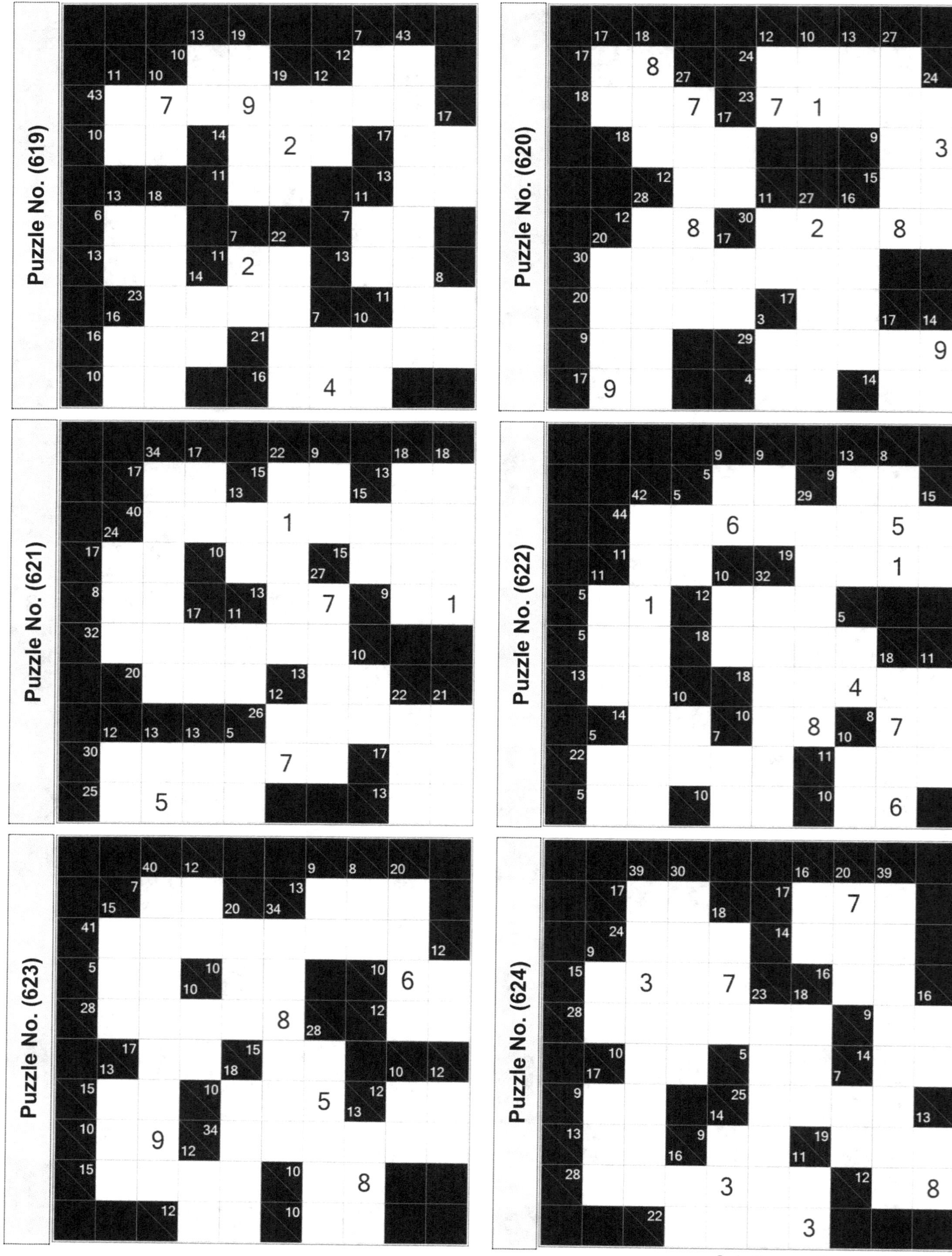

Puzzle No. (619)

Puzzle No. (620)

Puzzle No. (621)

Puzzle No. (622)

Puzzle No. (623)

Puzzle No. (624)

Solution on Pages (200-201)

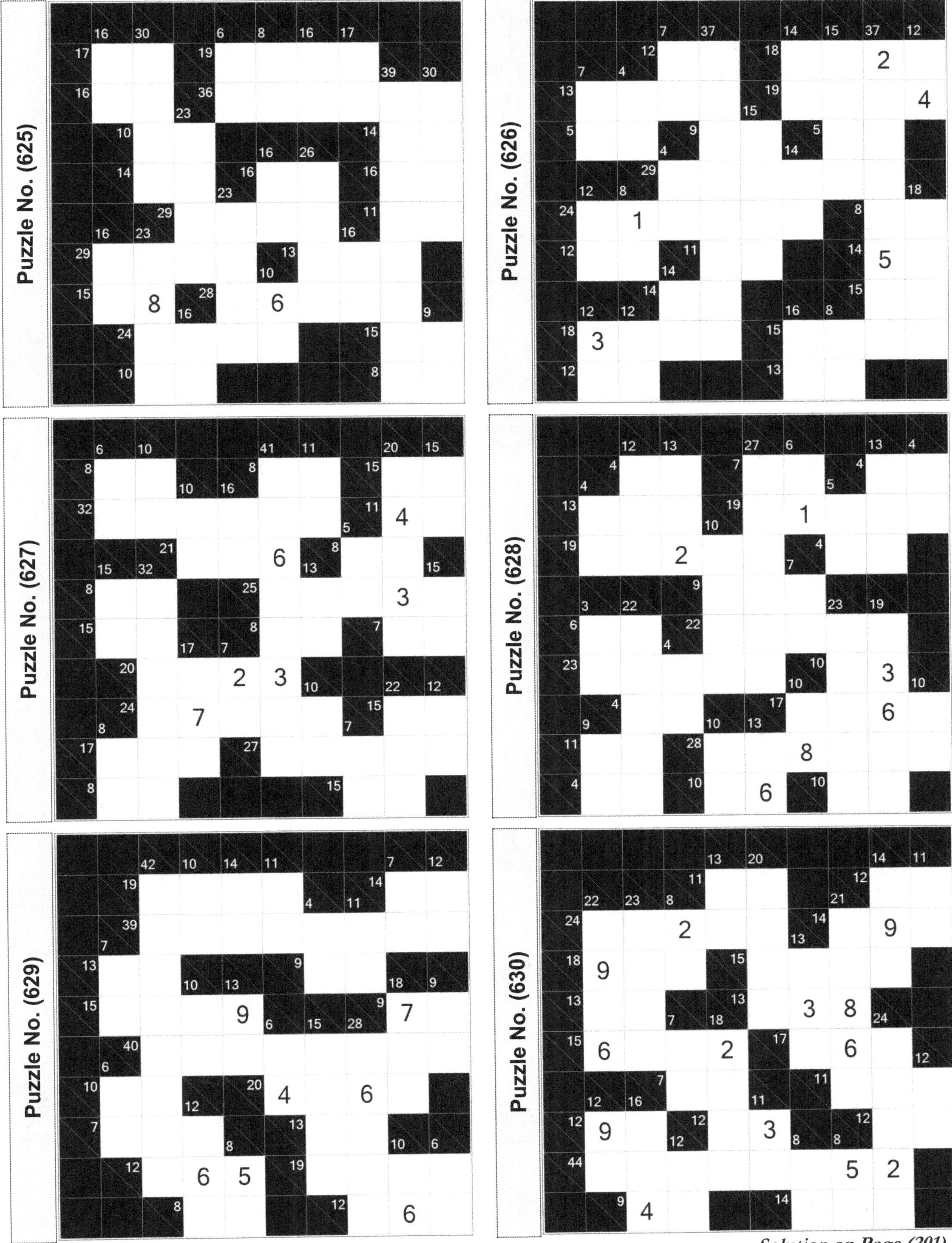

Puzzle No. (625)
Puzzle No. (626)
Puzzle No. (627)
Puzzle No. (628)
Puzzle No. (629)
Puzzle No. (630)
Solution on Page (201)

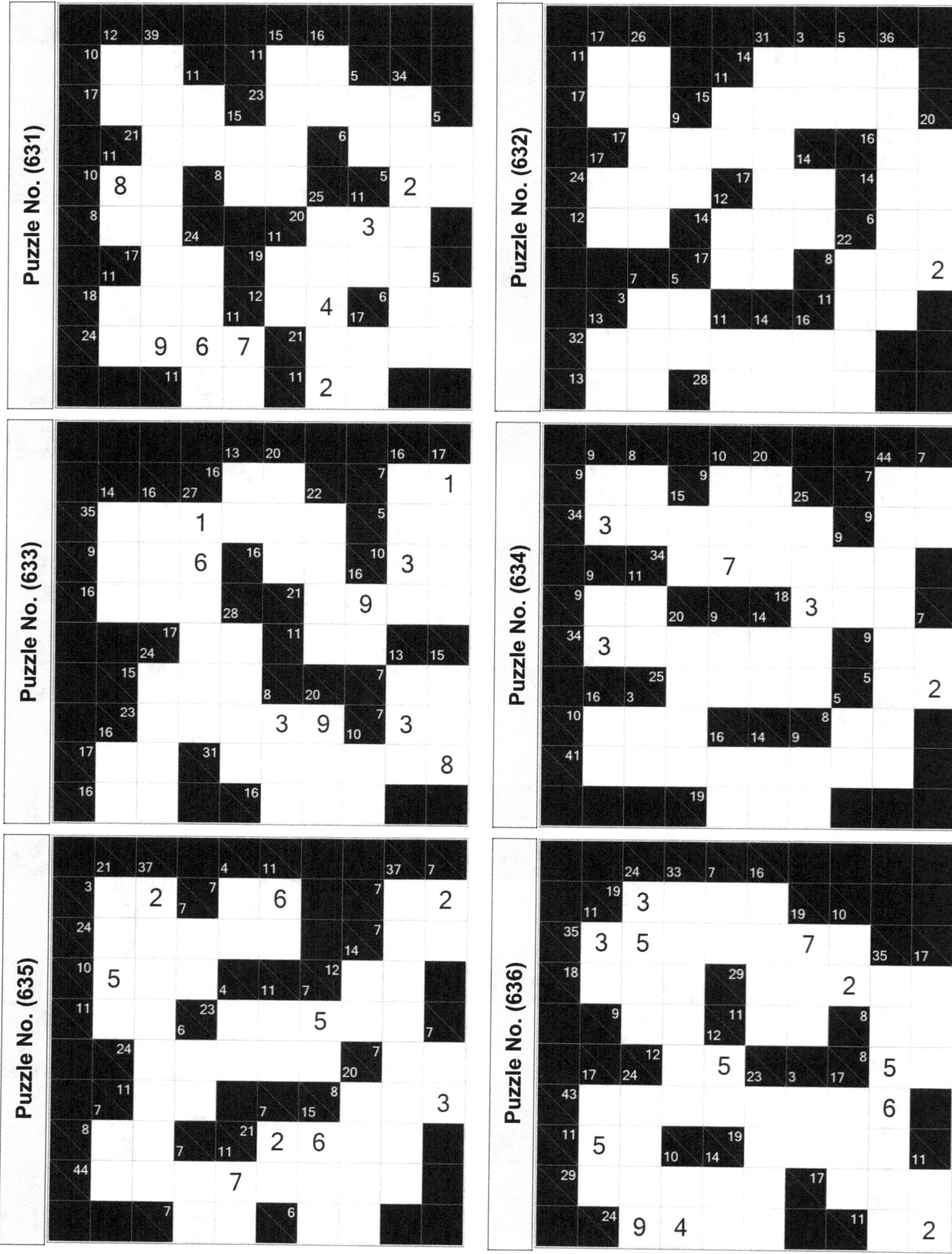

Solution on Page (201)

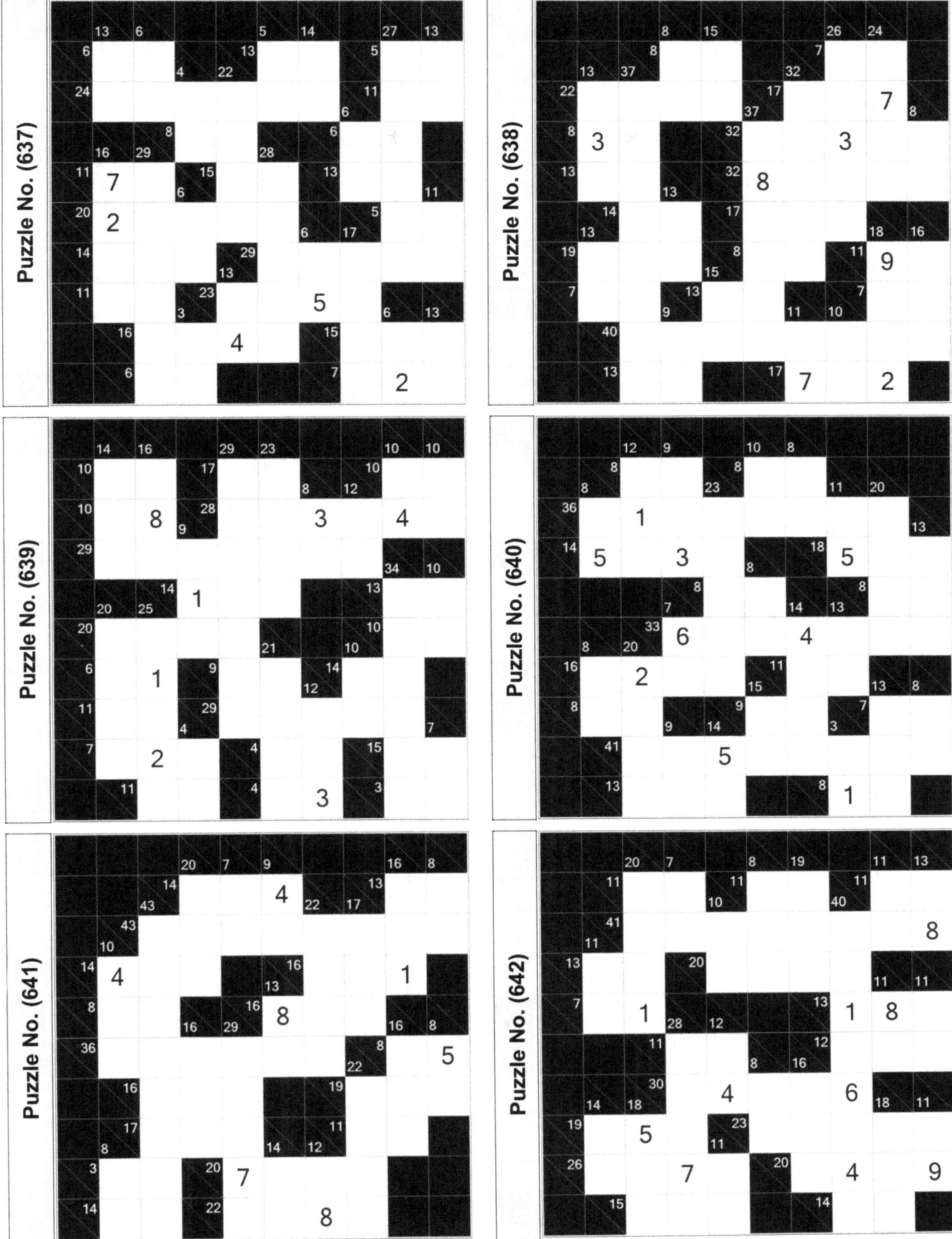

Puzzle No. (637)

Puzzle No. (638)

Puzzle No. (639)

Puzzle No. (640)

Puzzle No. (641)

Puzzle No. (642)

Solution on Pages (201-202)

(109)

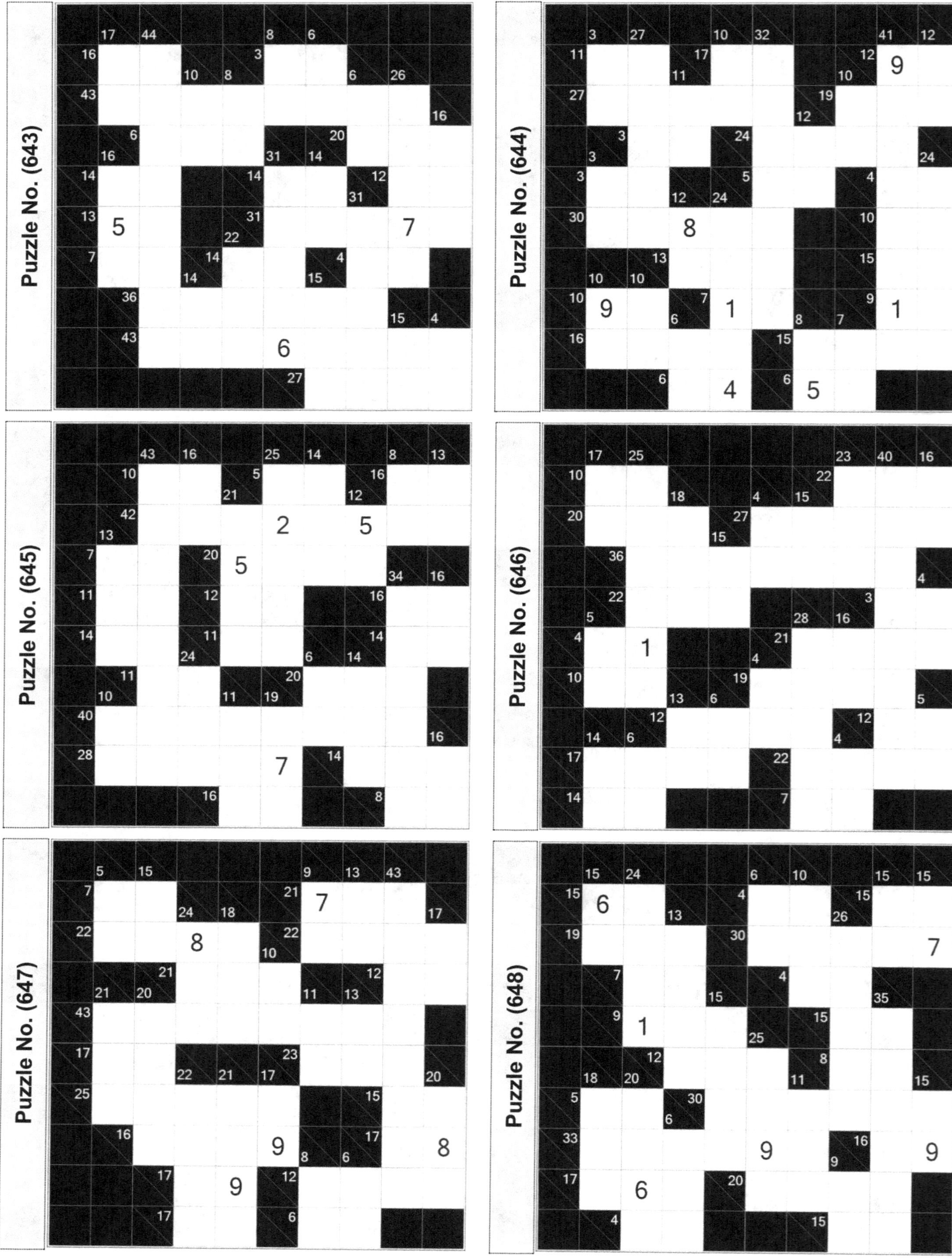

Puzzle No. (643)
Puzzle No. (644)
Puzzle No. (645)
Puzzle No. (646)
Puzzle No. (647)
Puzzle No. (648)

Solution on Page (202)

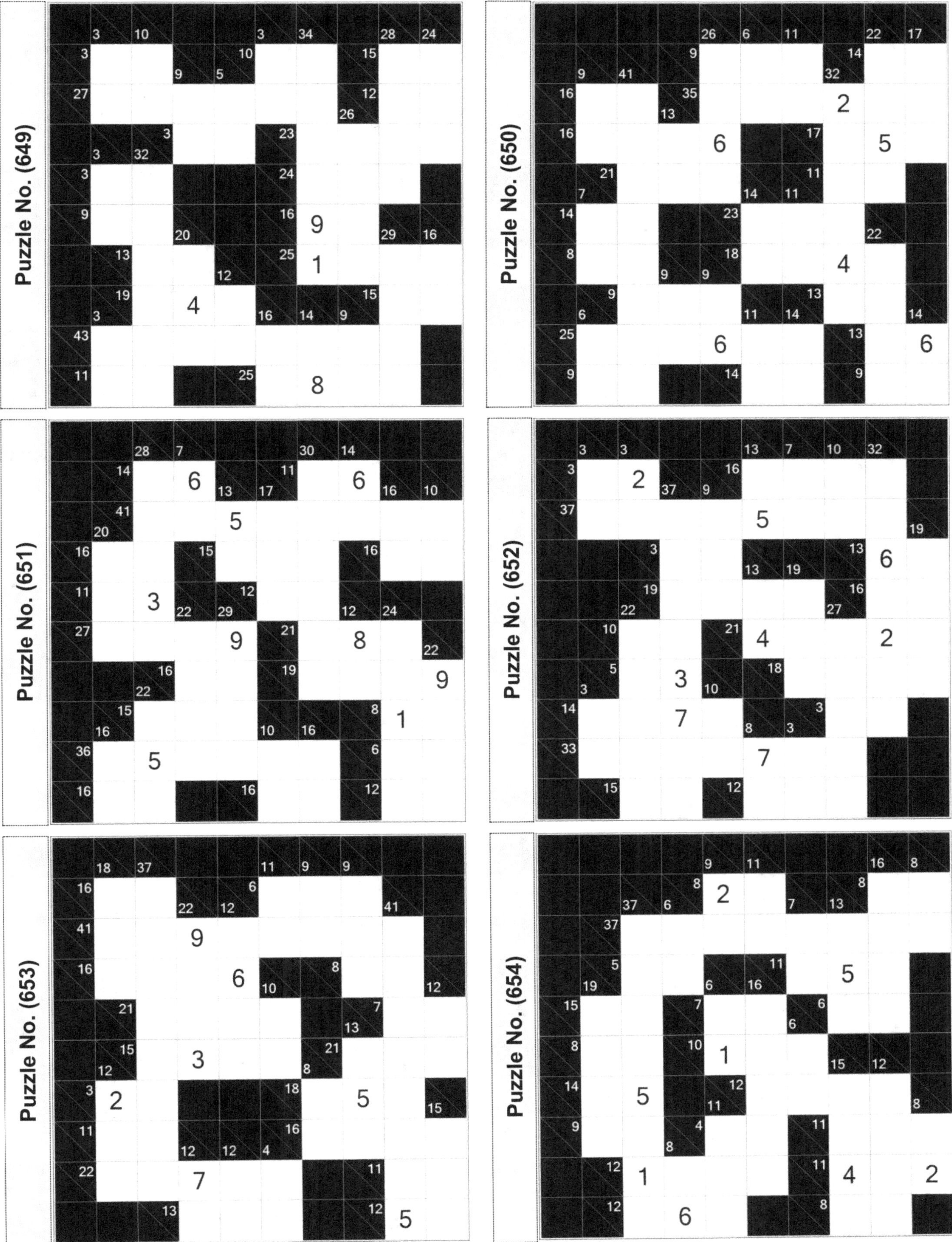

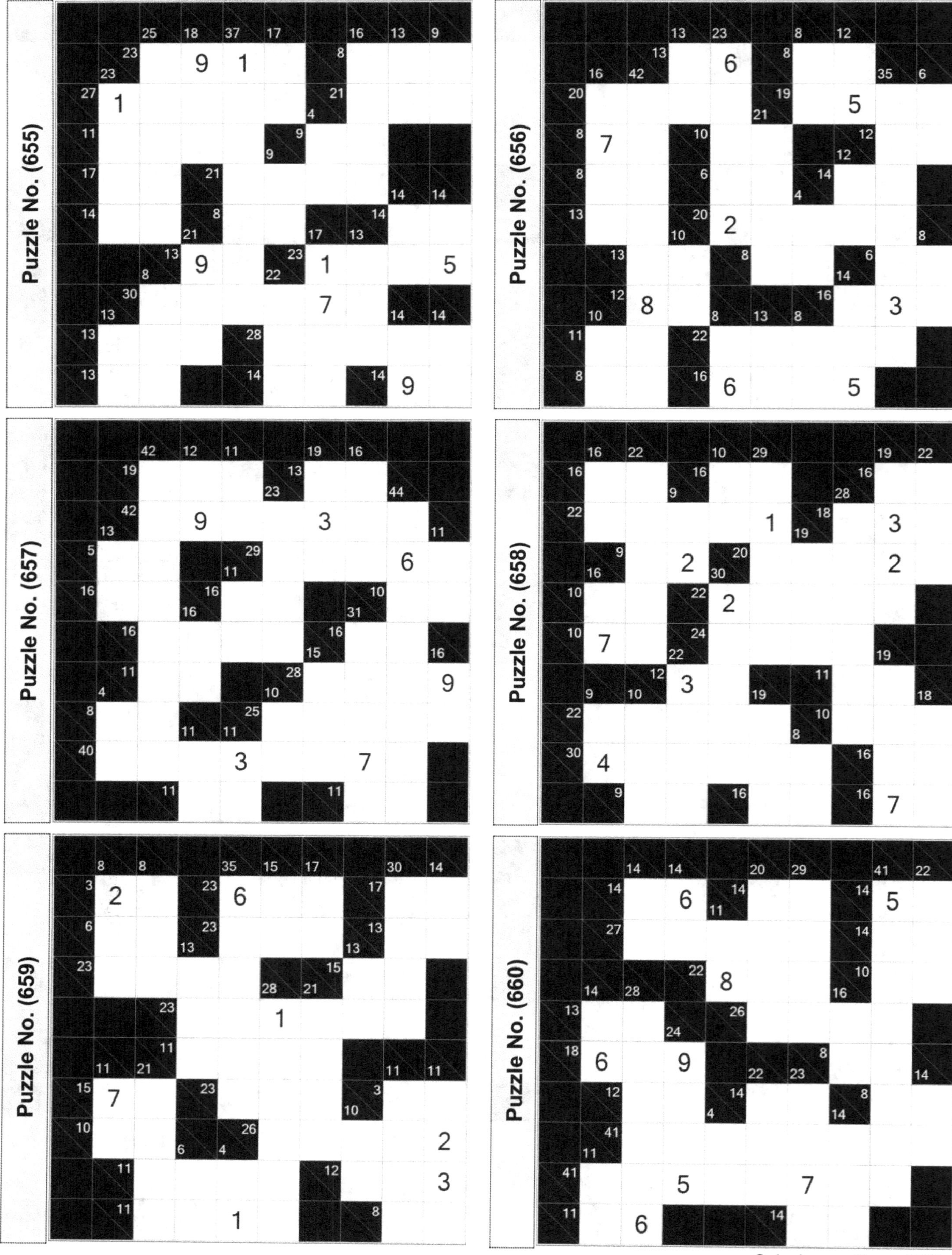

Solution on Page (202)

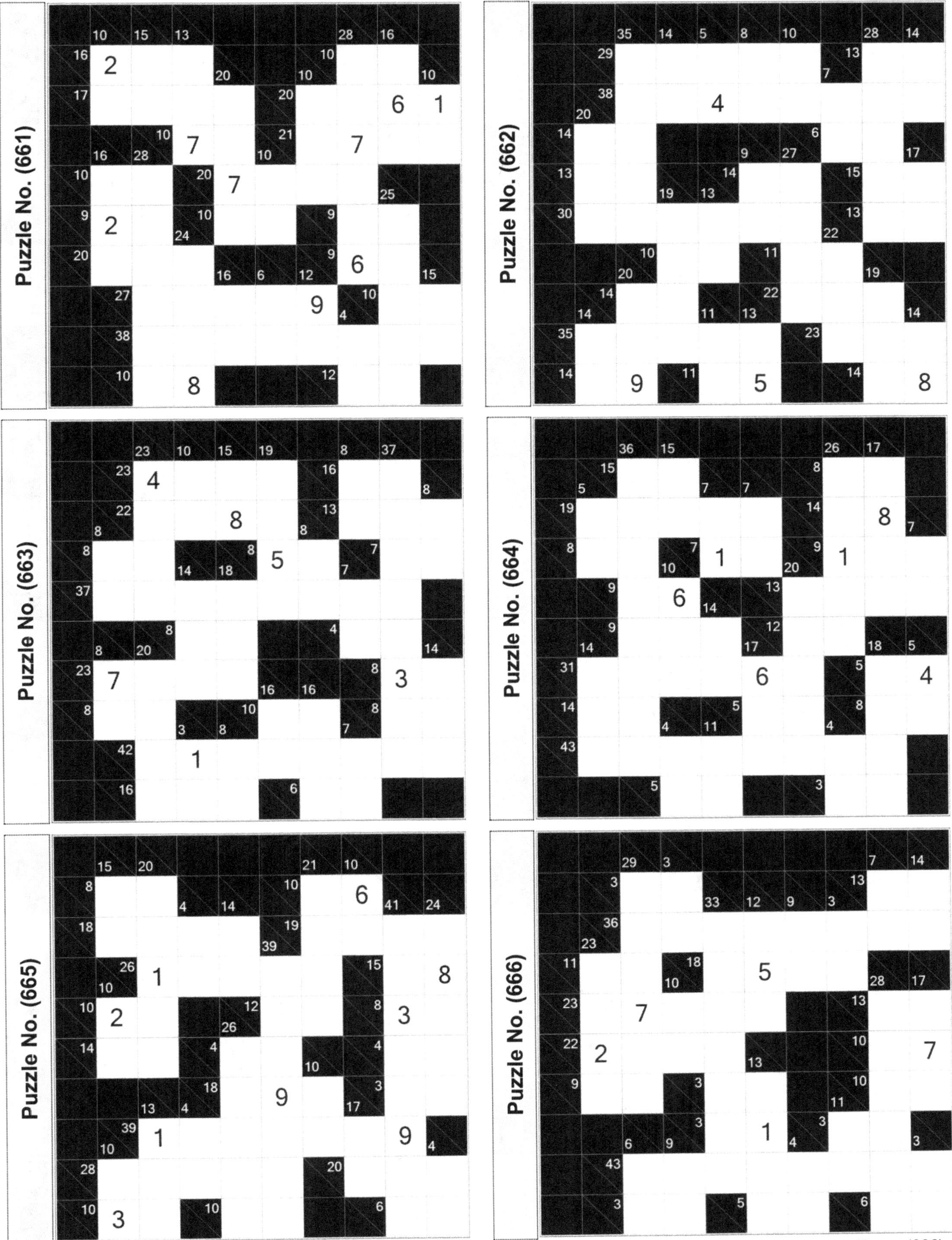

Puzzle No. (661)
Puzzle No. (662)
Puzzle No. (663)
Puzzle No. (664)
Puzzle No. (665)
Puzzle No. (666)
Solution on Page (203)

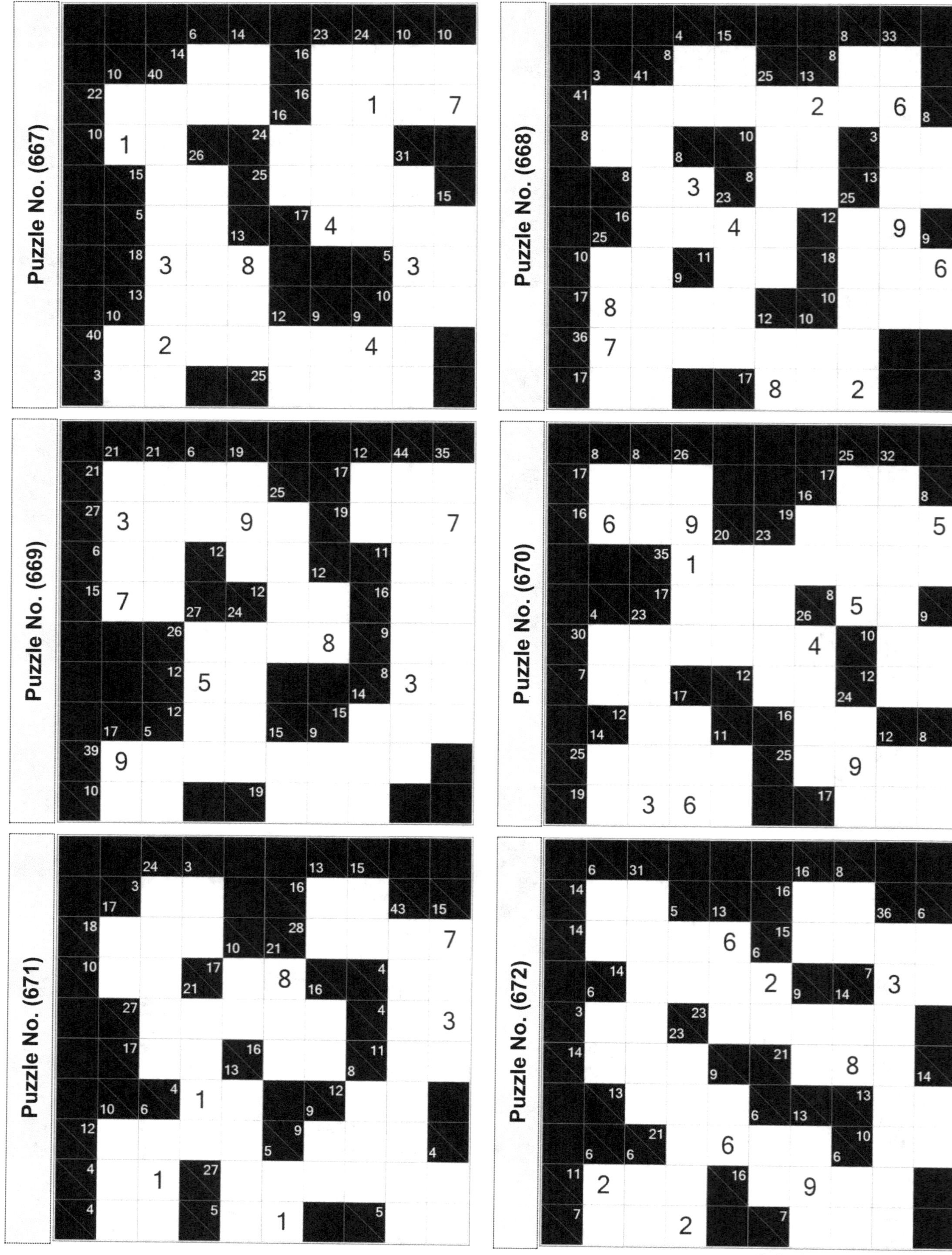

Puzzle No. (667)

Puzzle No. (668)

Puzzle No. (669)

Puzzle No. (670)

Puzzle No. (671)

Puzzle No. (672)

Solution on Page (203)

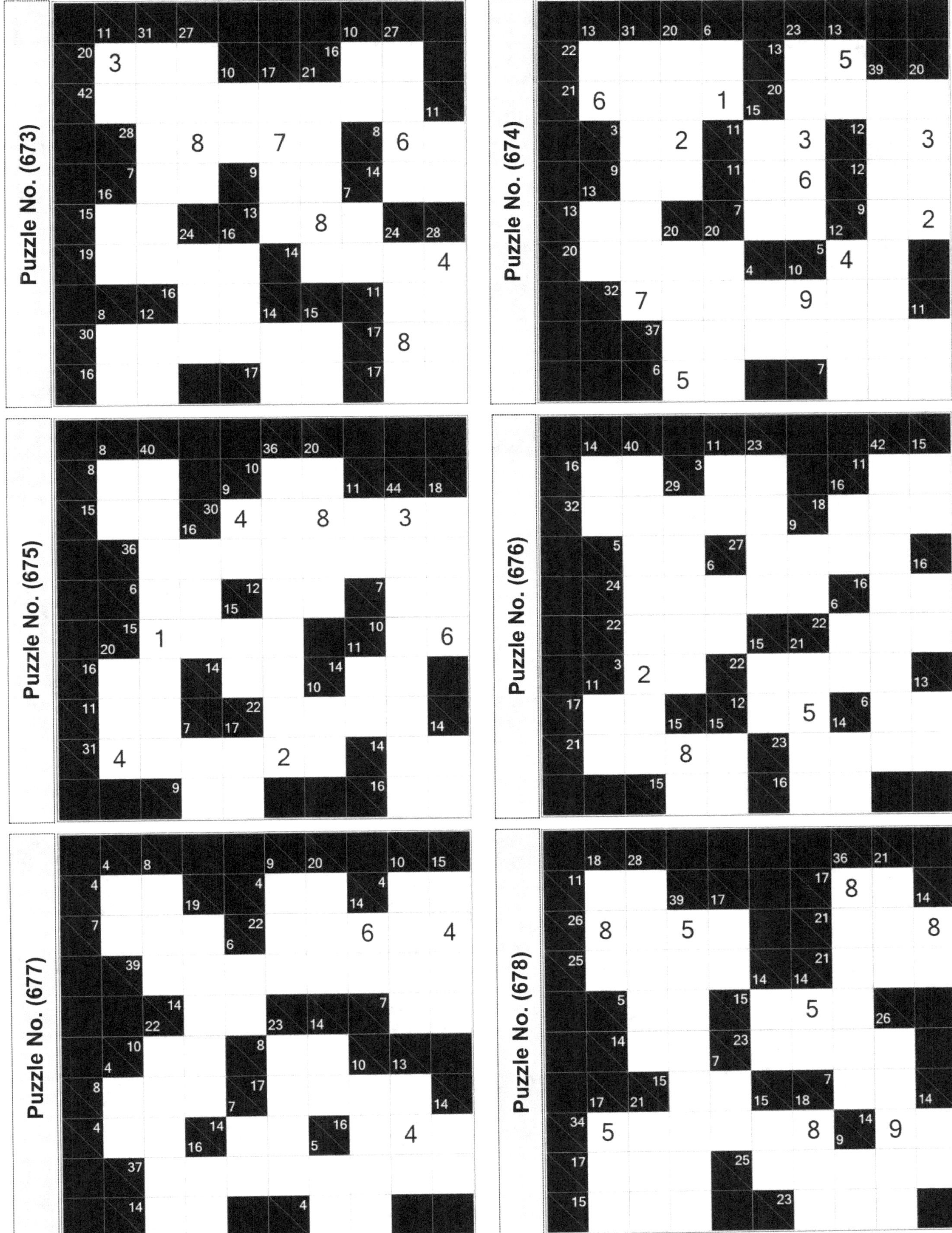

(115)

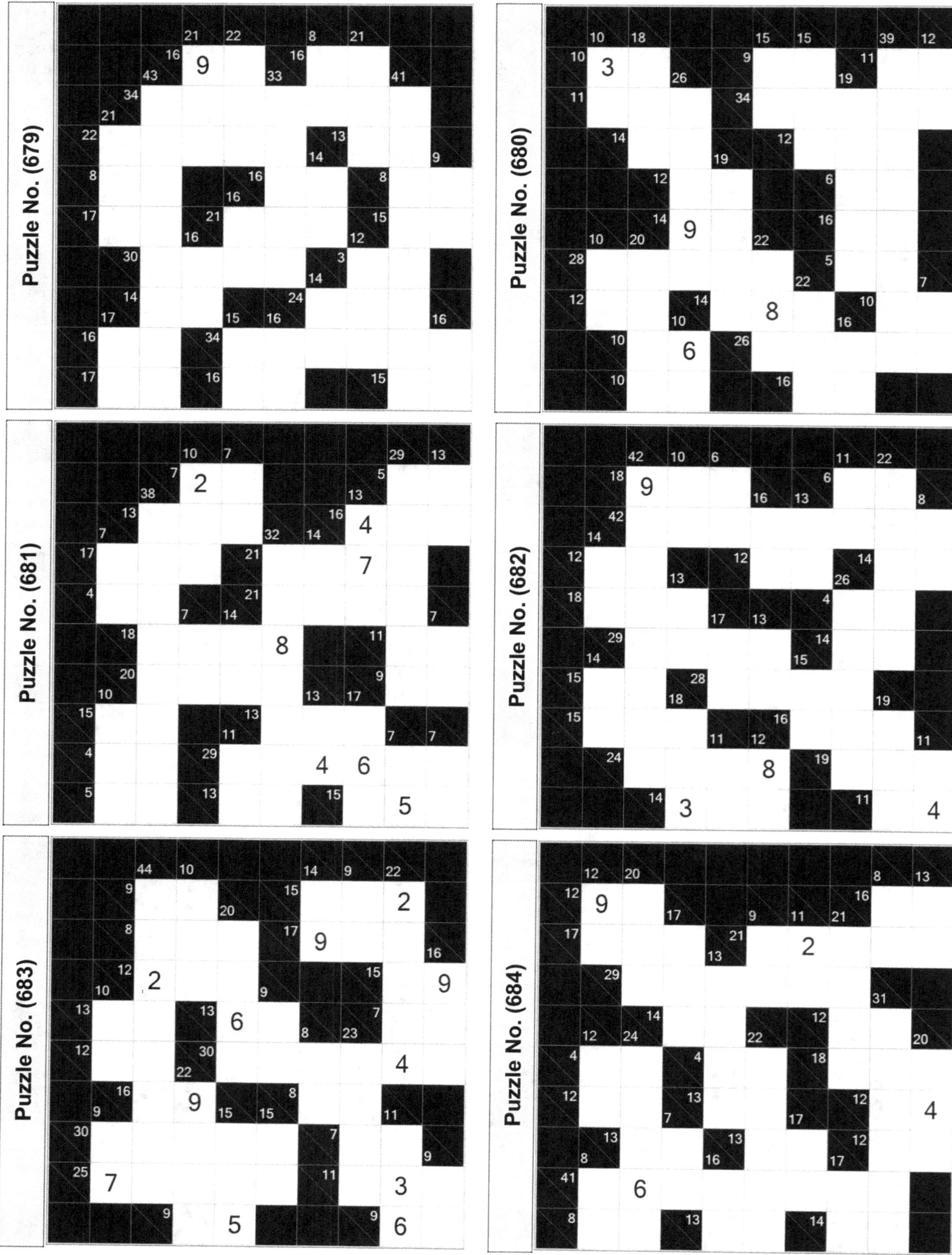

Solution on Page (203-204)

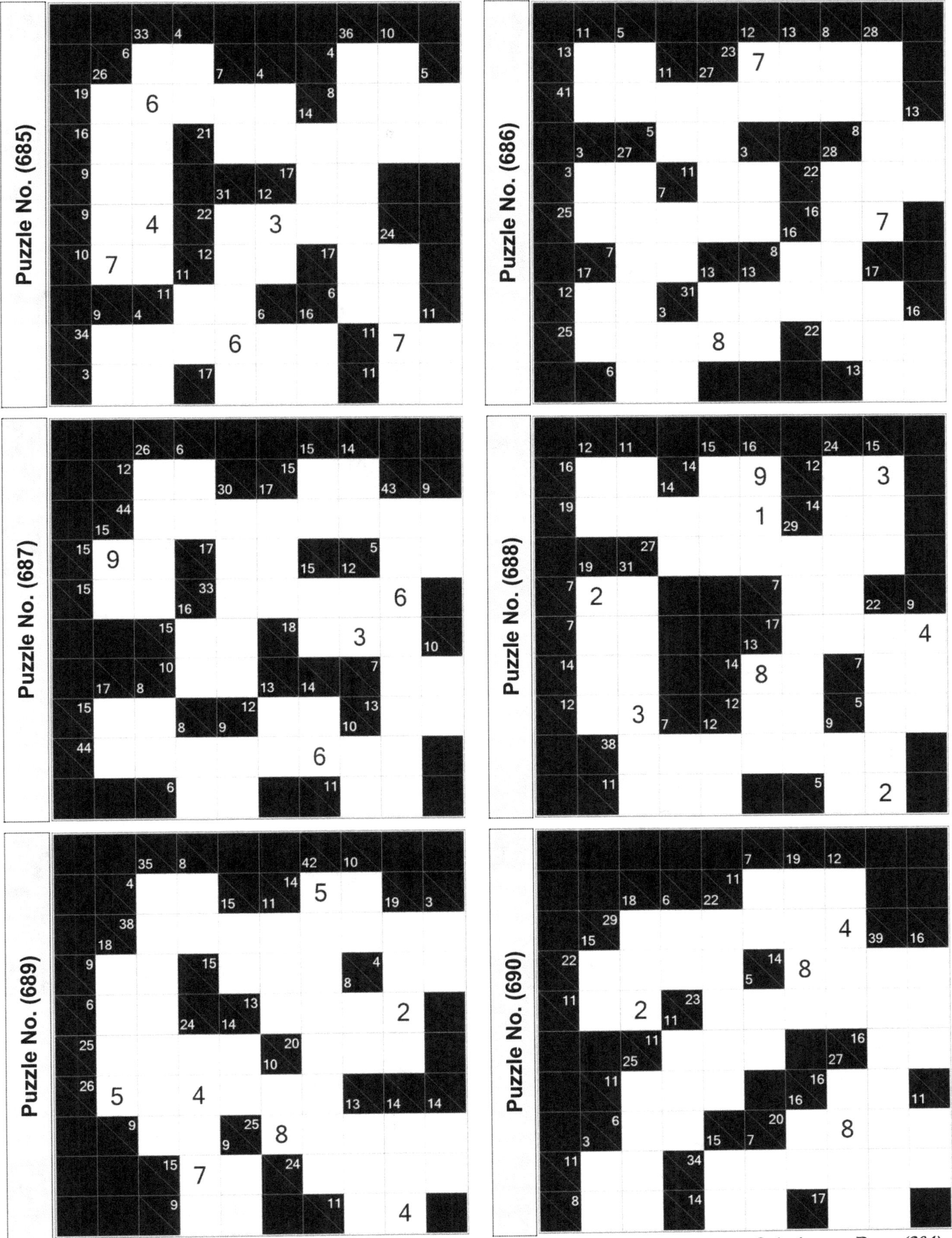

Puzzle No. (685)

Puzzle No. (686)

Puzzle No. (687)

Puzzle No. (688)

Puzzle No. (689)

Puzzle No. (690)

Solution on Page (204)

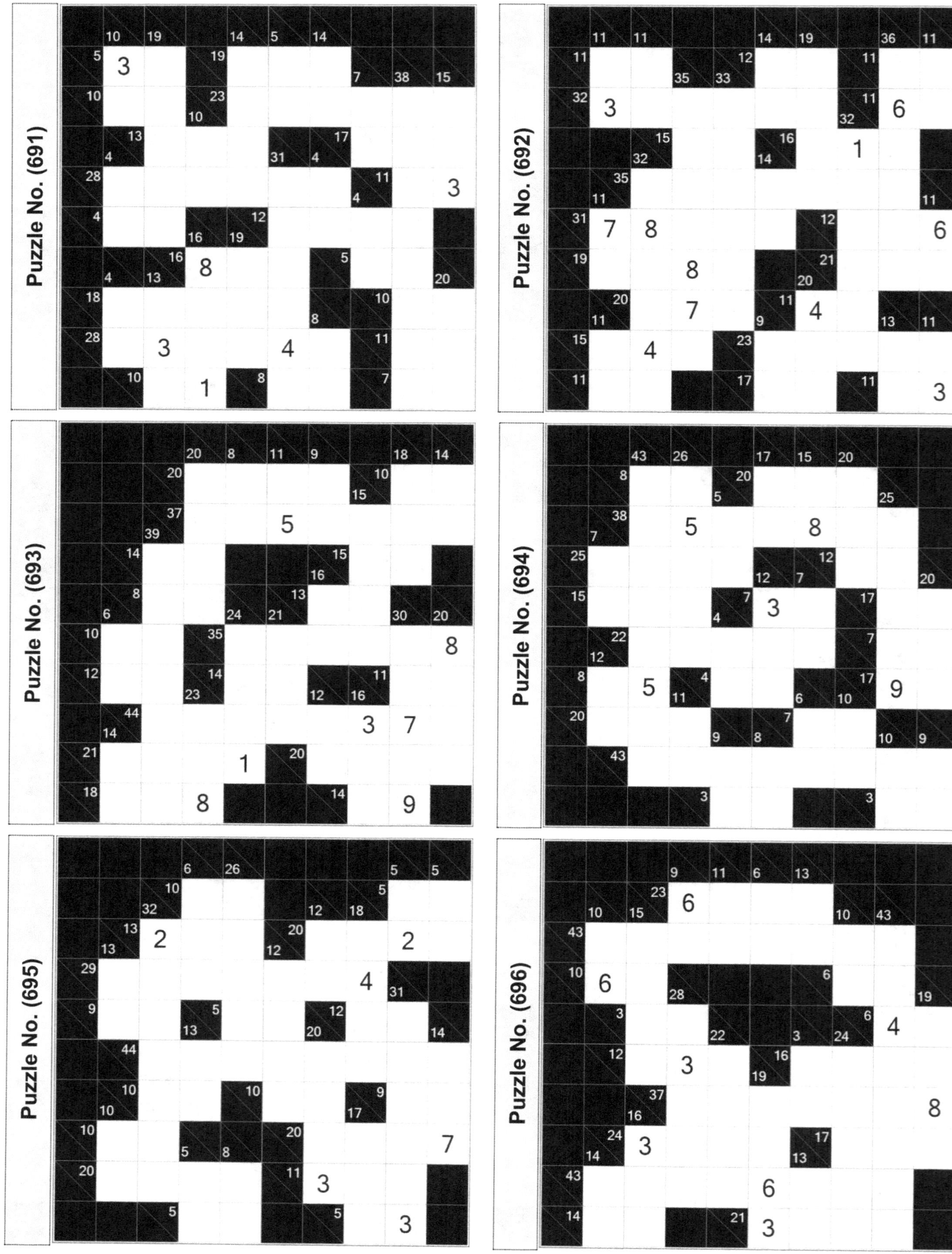

Puzzle No. (691)
Puzzle No. (692)
Puzzle No. (693)
Puzzle No. (694)
Puzzle No. (695)
Puzzle No. (696)
Solution on Page (204)
(118)

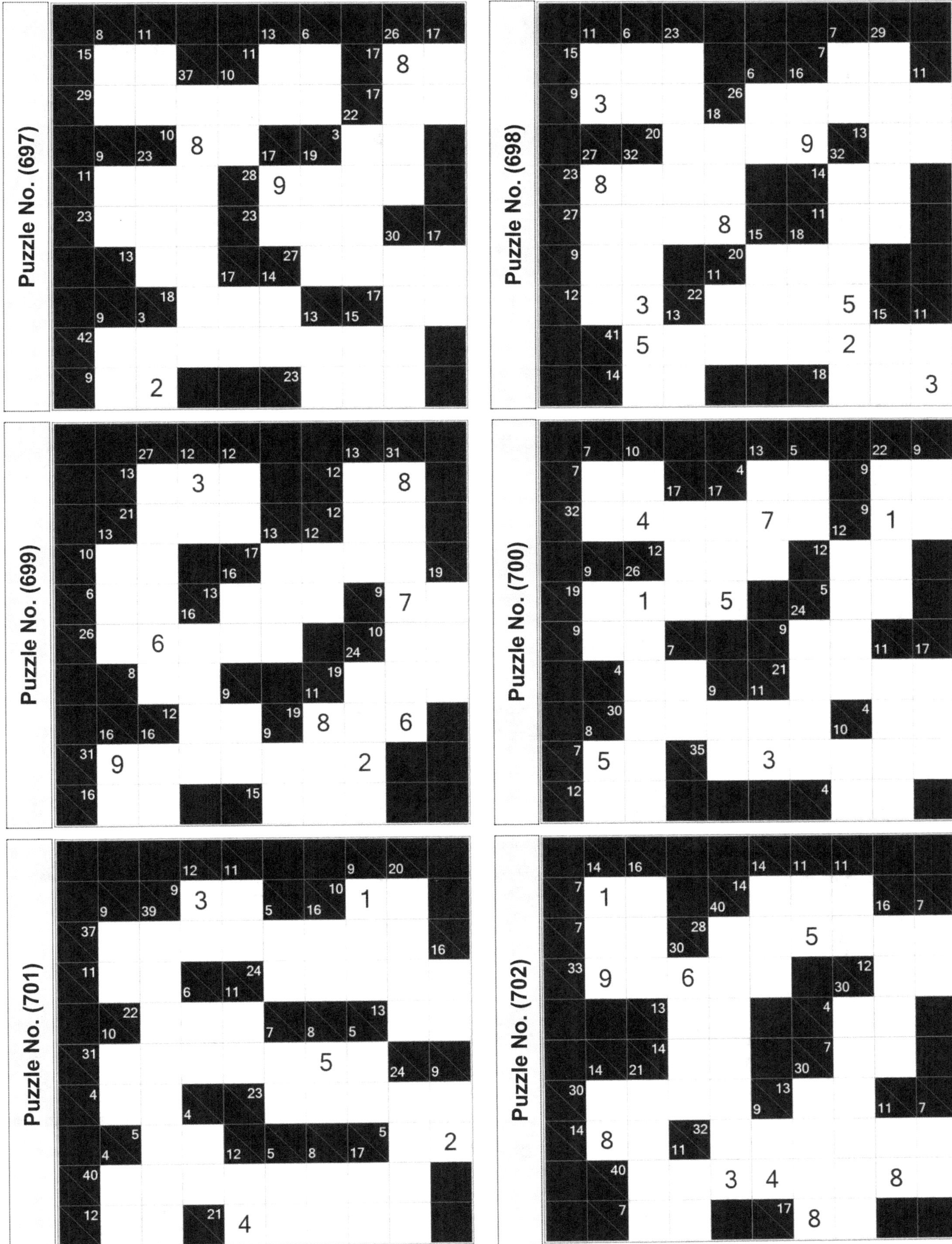

Puzzle No. (697)
Puzzle No. (698)
Puzzle No. (699)
Puzzle No. (700)
Puzzle No. (701)
Puzzle No. (702)
Solution on Pages (204-205)

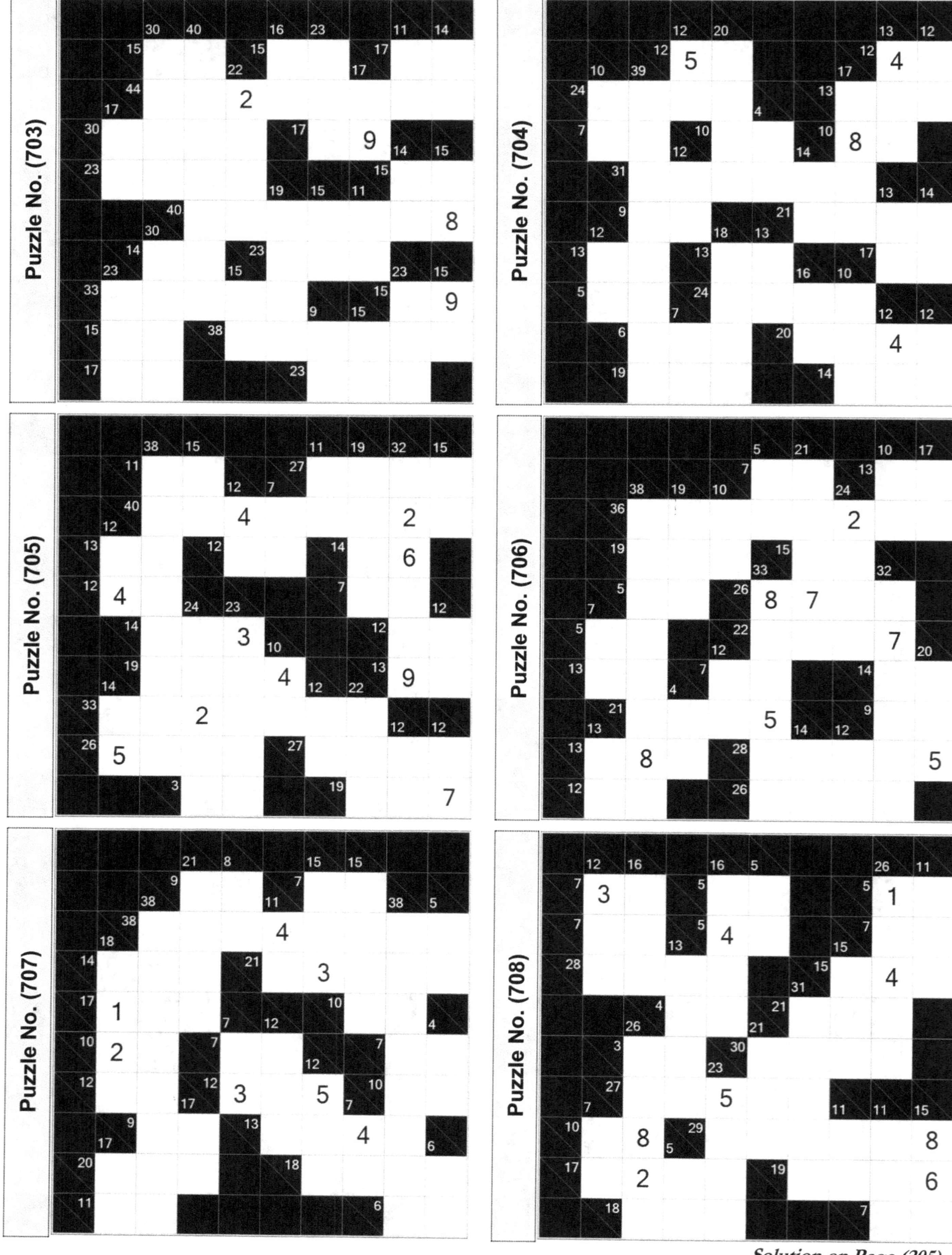

Puzzle No. (703)

Puzzle No. (704)

Puzzle No. (705)

Puzzle No. (706)

Puzzle No. (707)

Puzzle No. (708)

Solution on Page (205)

(120)

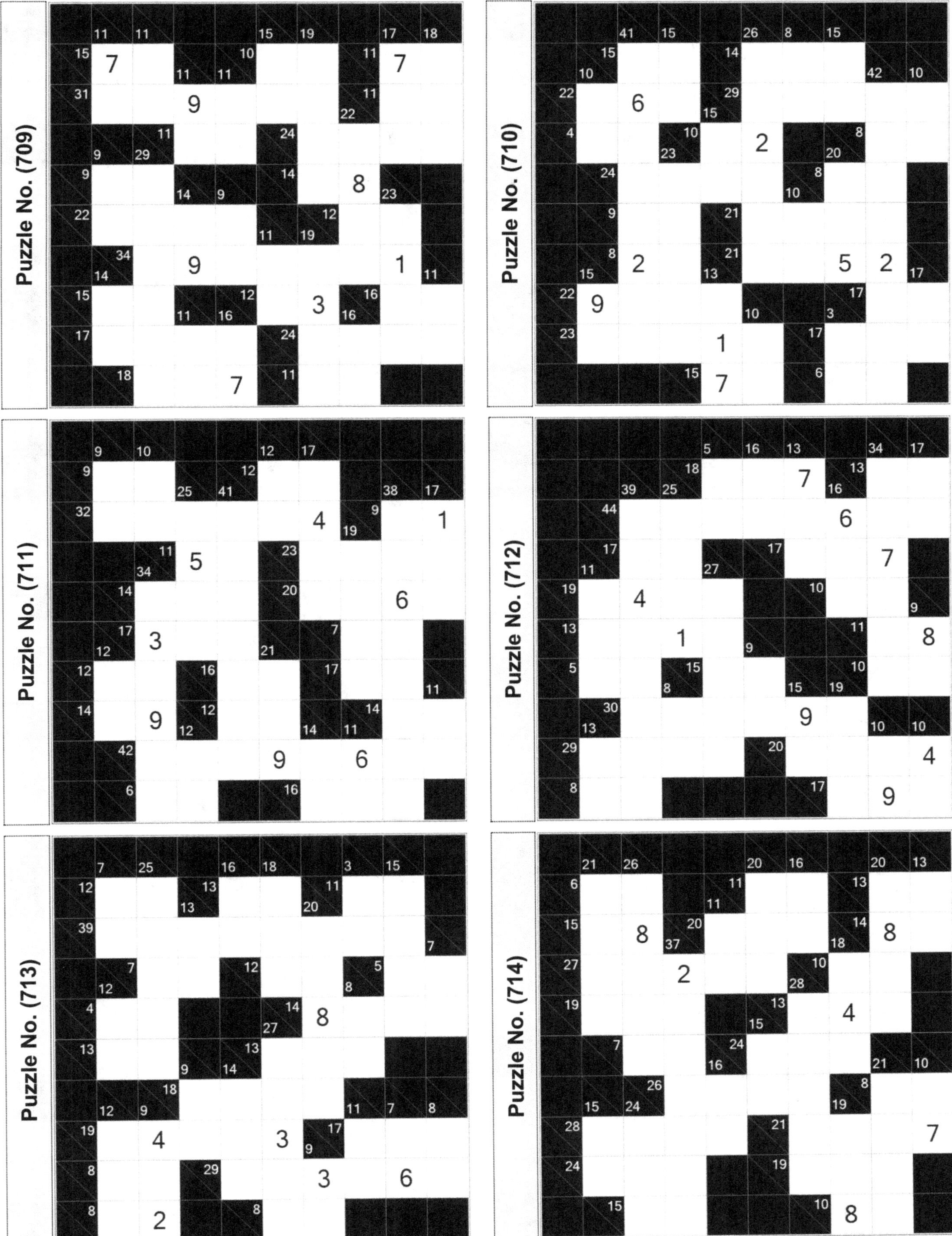

Solution on Page (205)

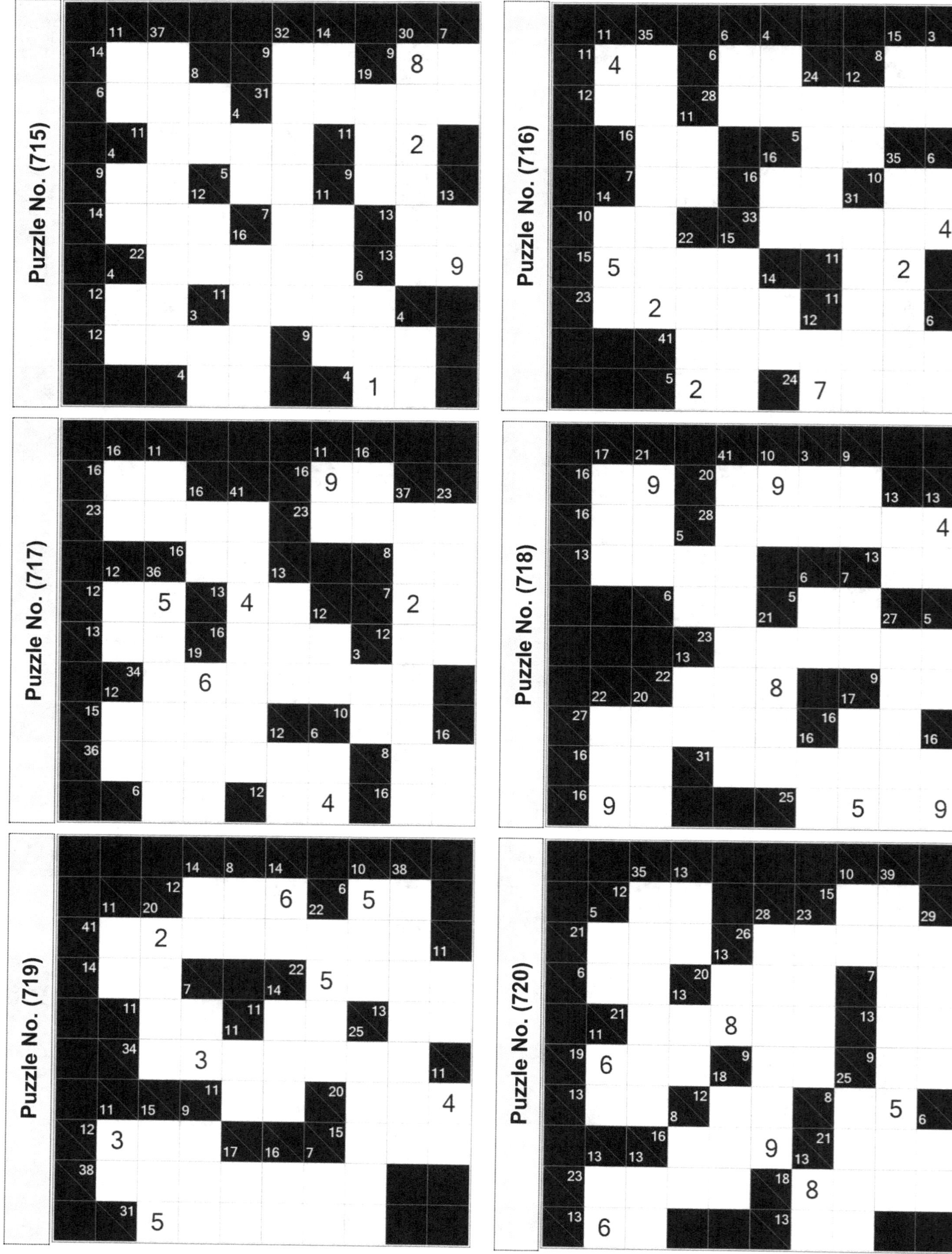

Puzzle No. (715)

Puzzle No. (716)

Puzzle No. (717)

Puzzle No. (718)

Puzzle No. (719)

Puzzle No. (720)

Solution on Page (205)

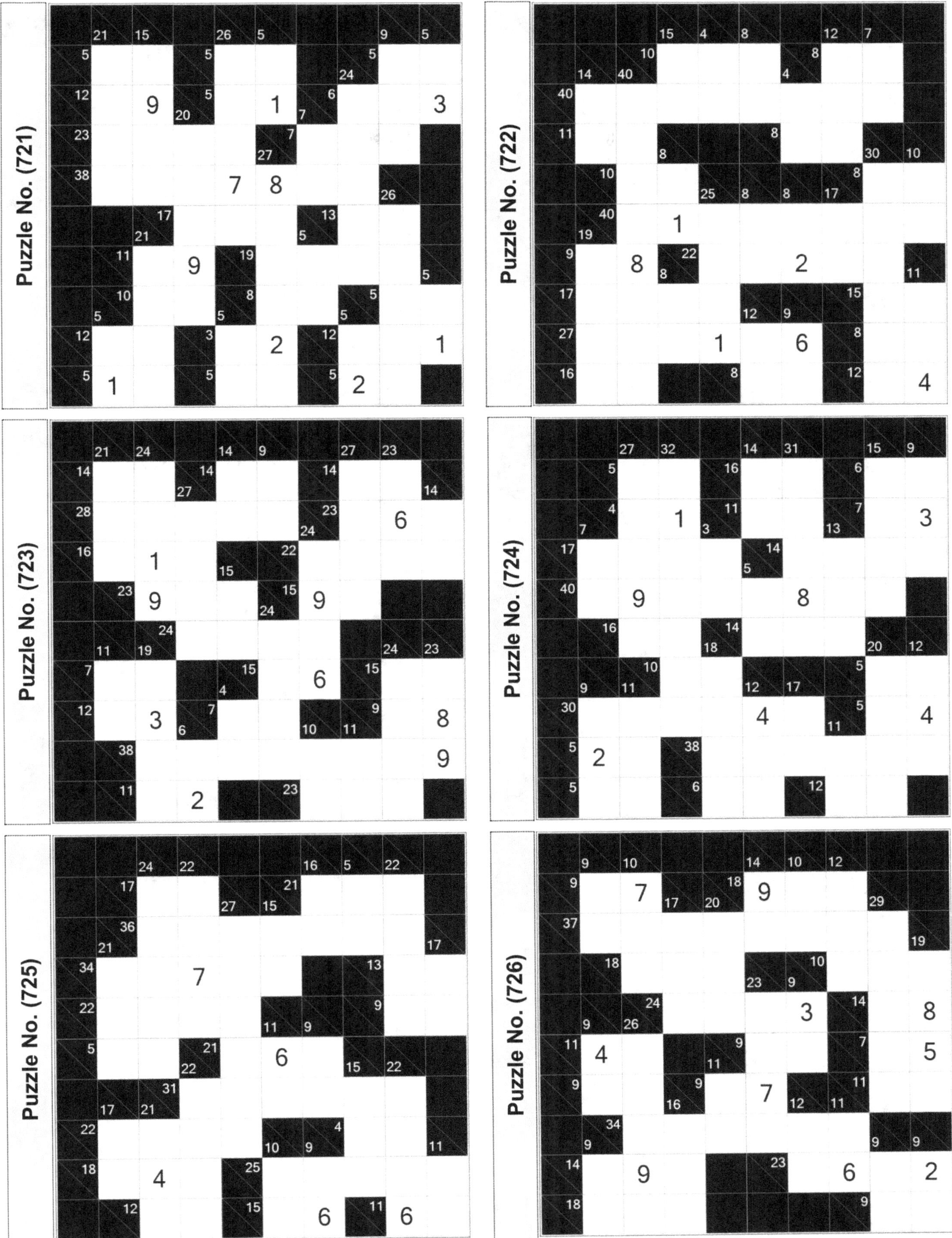

Puzzle No. (721)
Puzzle No. (722)
Puzzle No. (723)
Puzzle No. (724)
Puzzle No. (725)
Puzzle No. (726)
Solution on Page (206)

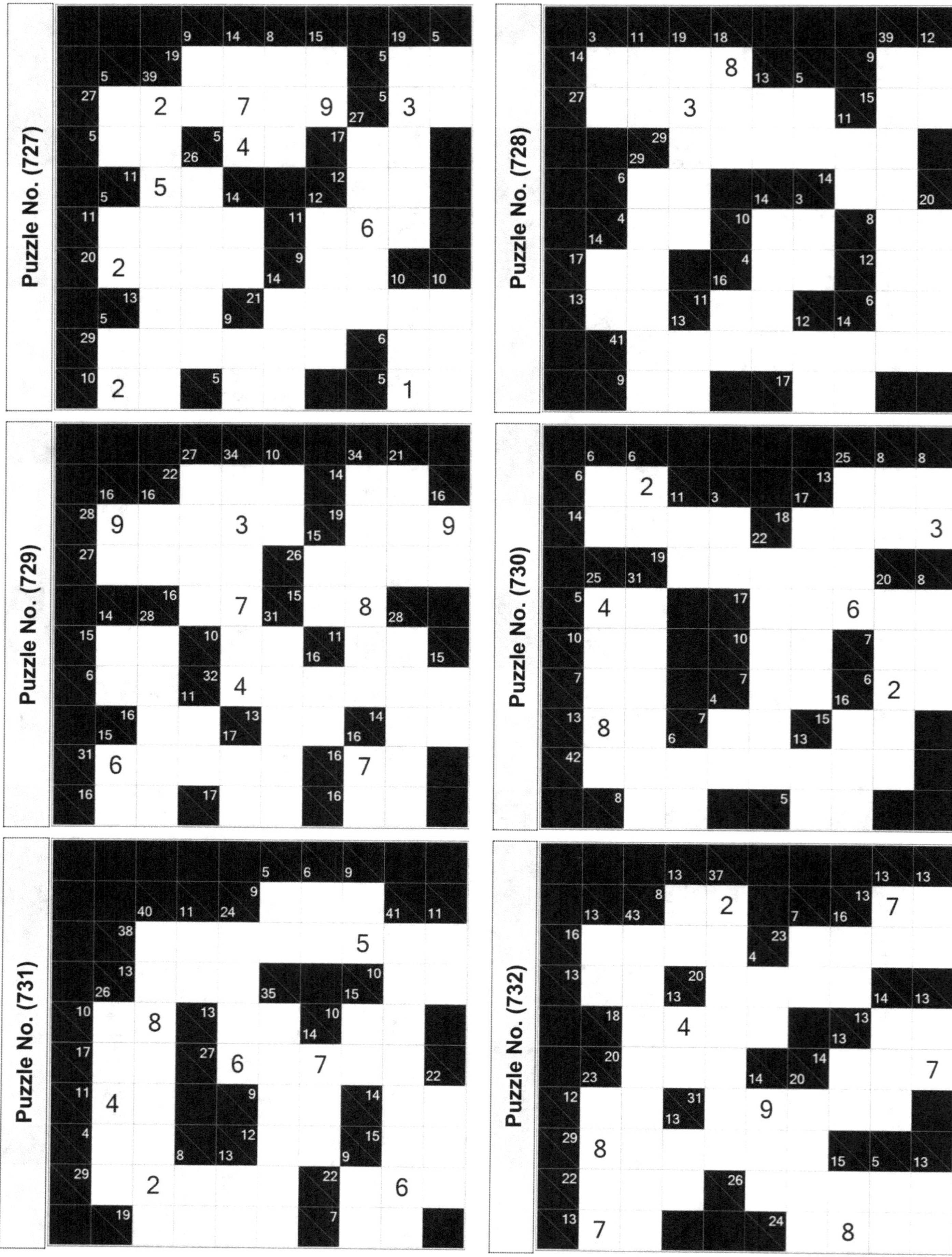

Puzzle No. (727)
Puzzle No. (728)
Puzzle No. (729)
Puzzle No. (730)
Puzzle No. (731)
Puzzle No. (732)

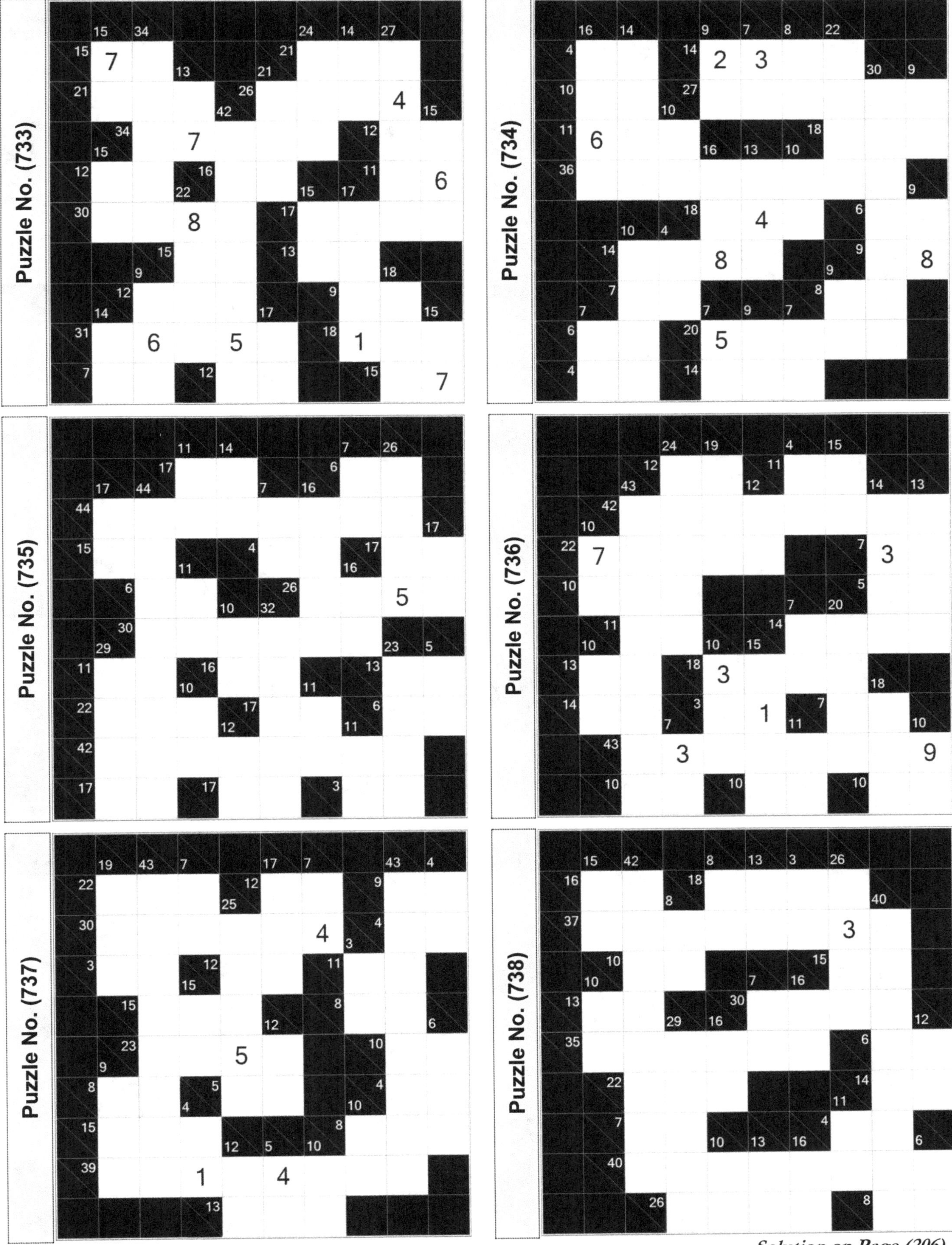

Puzzle No. (733)

Puzzle No. (734)

Puzzle No. (735)

Puzzle No. (736)

Puzzle No. (737)

Puzzle No. (738)

Solution on Page (206)

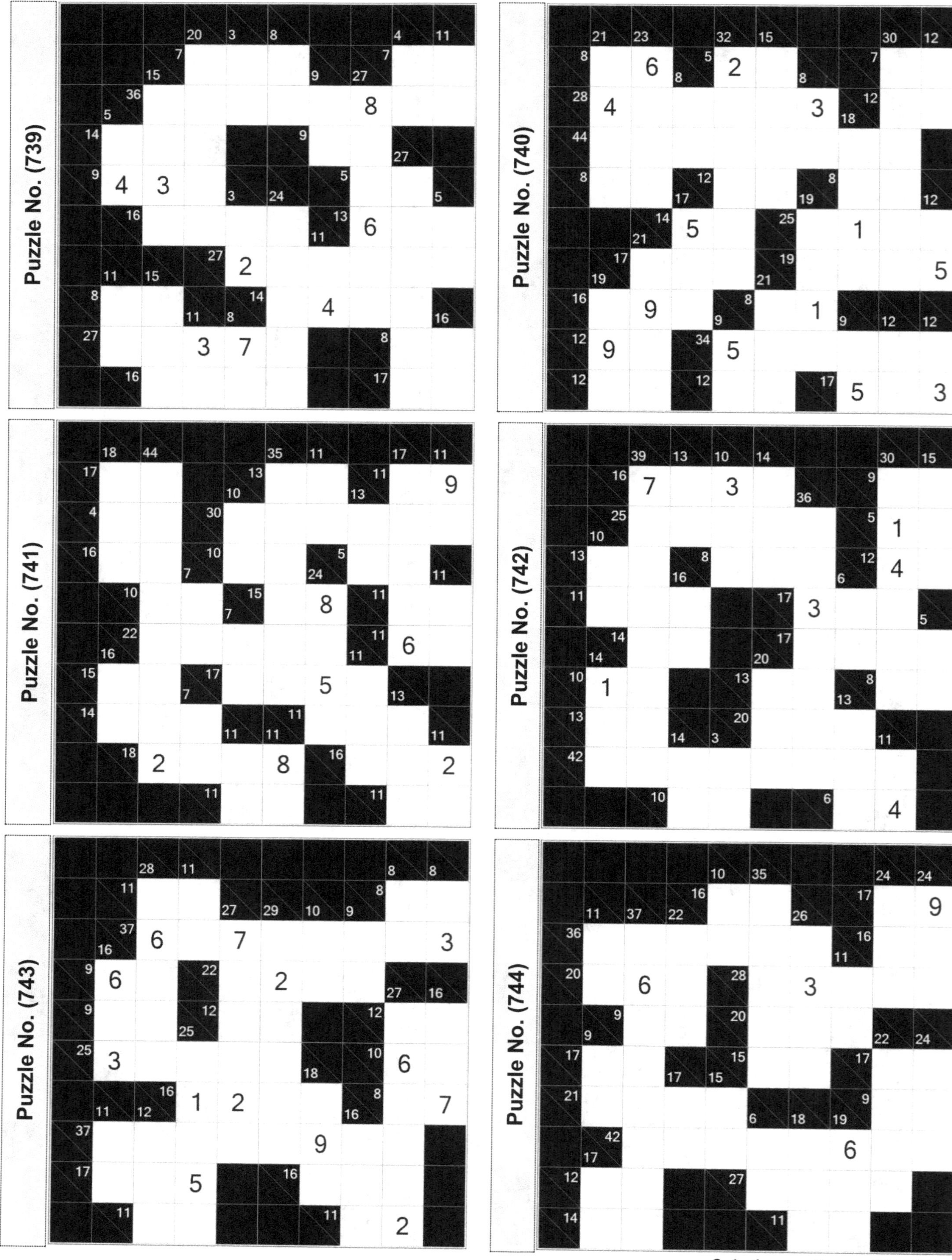

Solution on Pages (206-207)

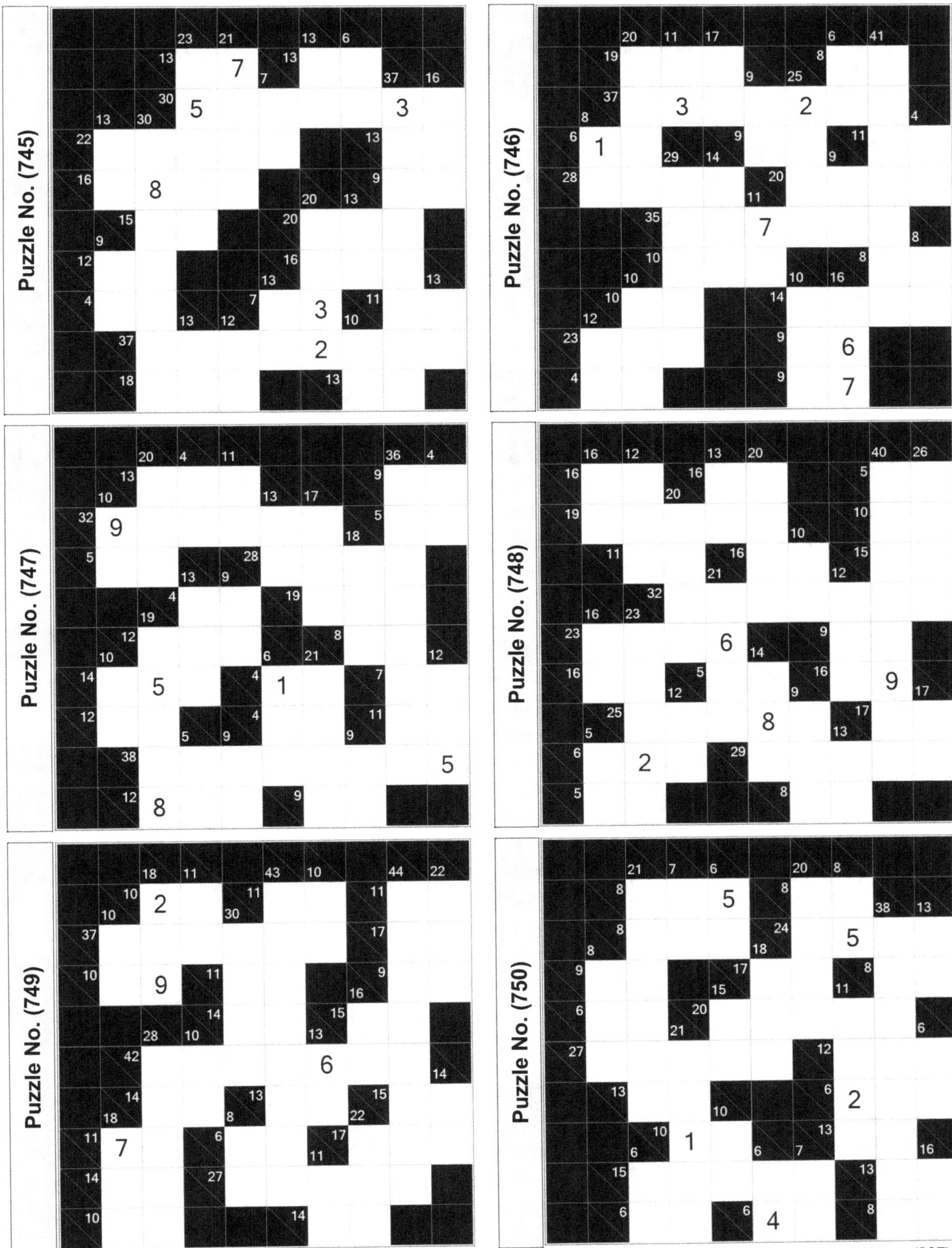

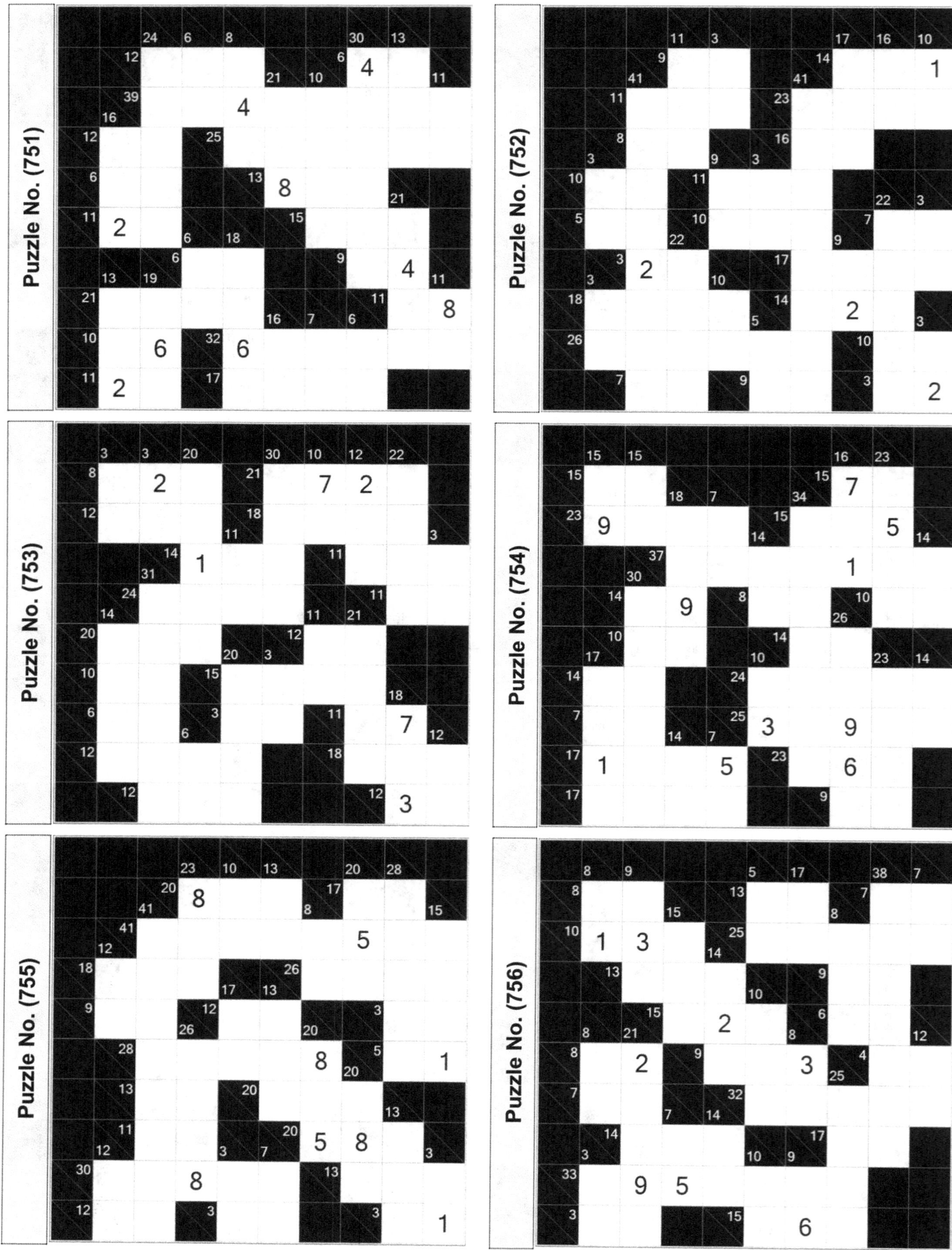

Solution on Page (207)

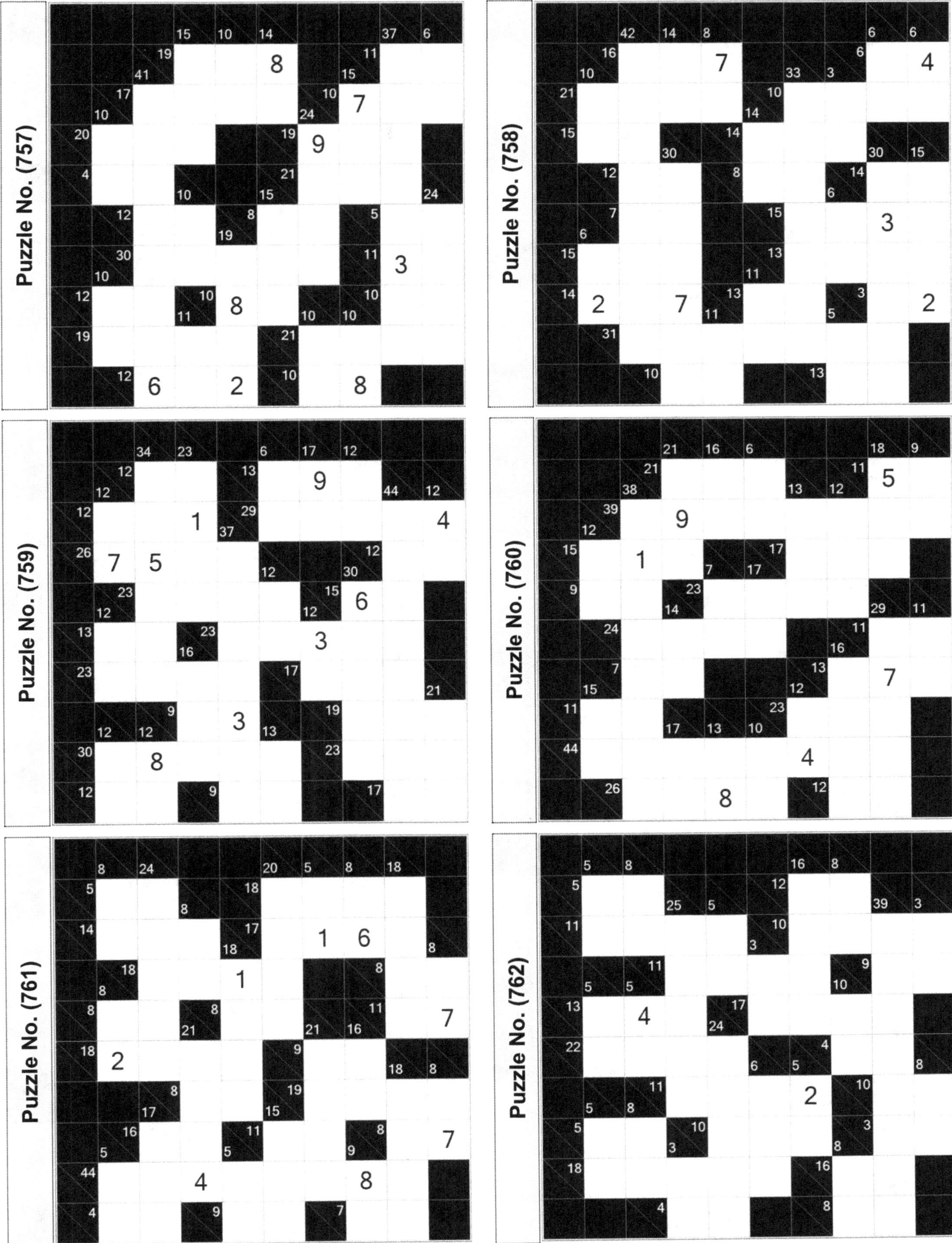

Puzzle No. (757)

Puzzle No. (758)

Puzzle No. (759)

Puzzle No. (760)

Puzzle No. (761)

Puzzle No. (762)

Solution on Pages (207-208)

(129)

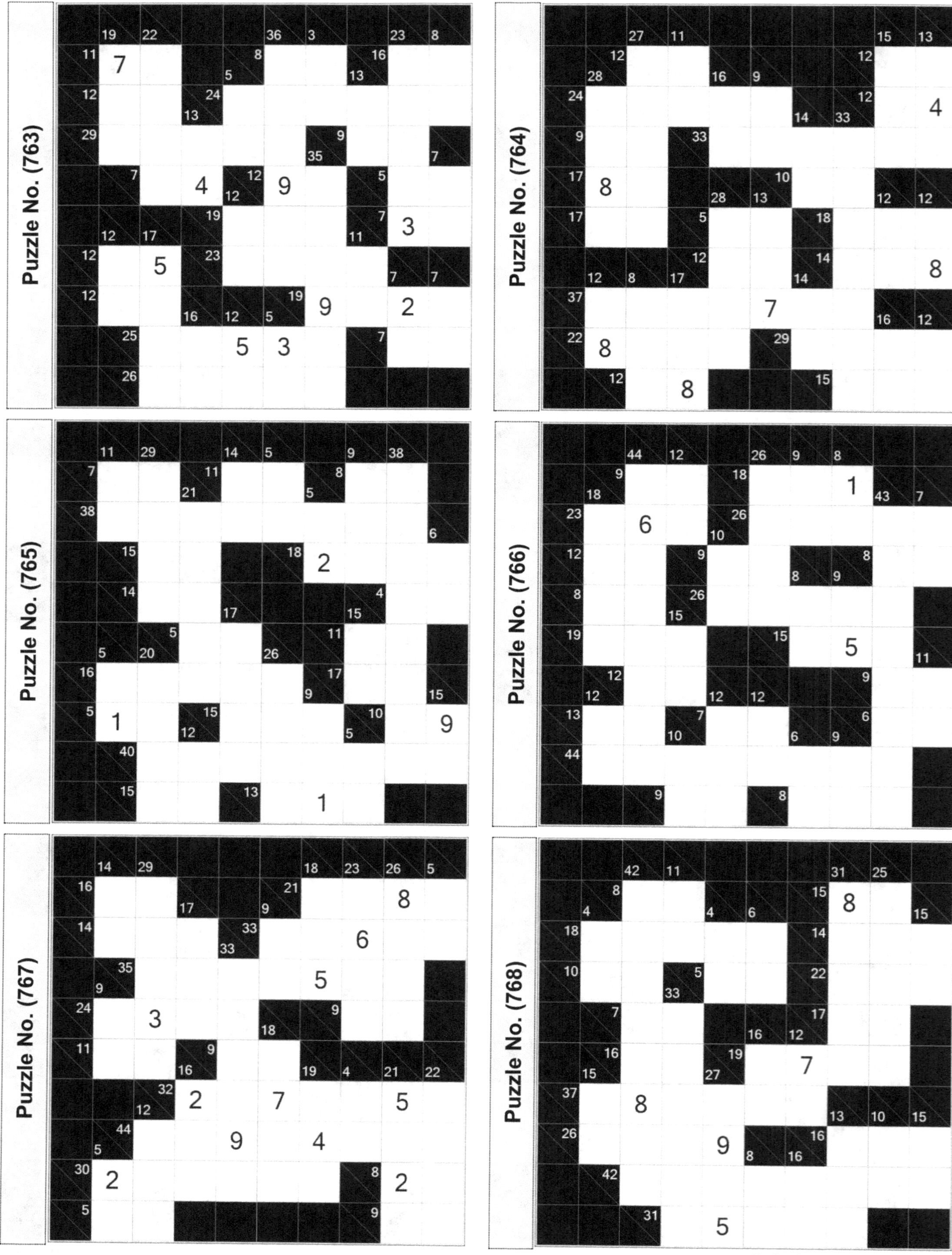

Solution on Page (208)

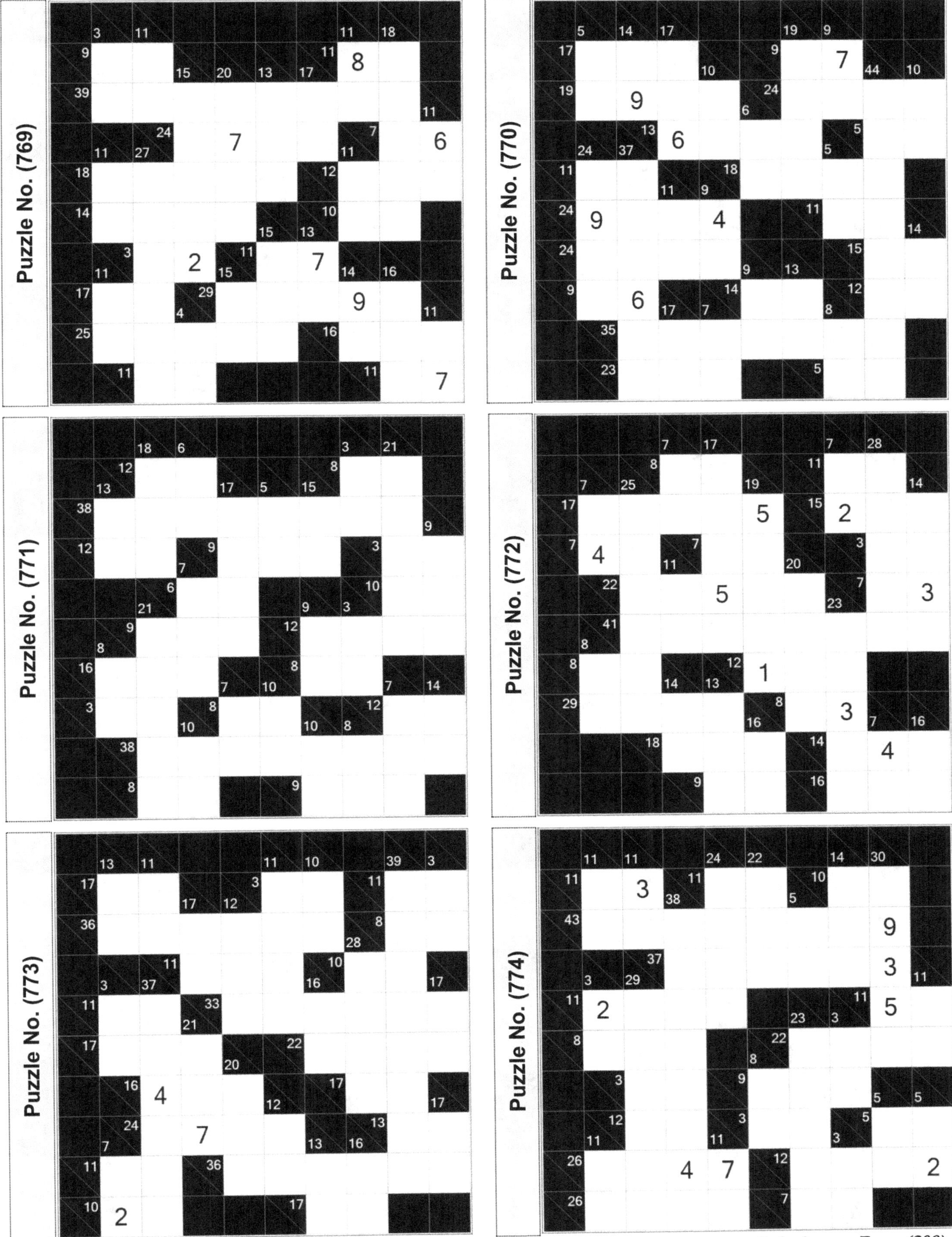

Puzzle No. (769)

Puzzle No. (770)

Puzzle No. (771)

Puzzle No. (772)

Puzzle No. (773)

Puzzle No. (774)

Solution on Page (208)

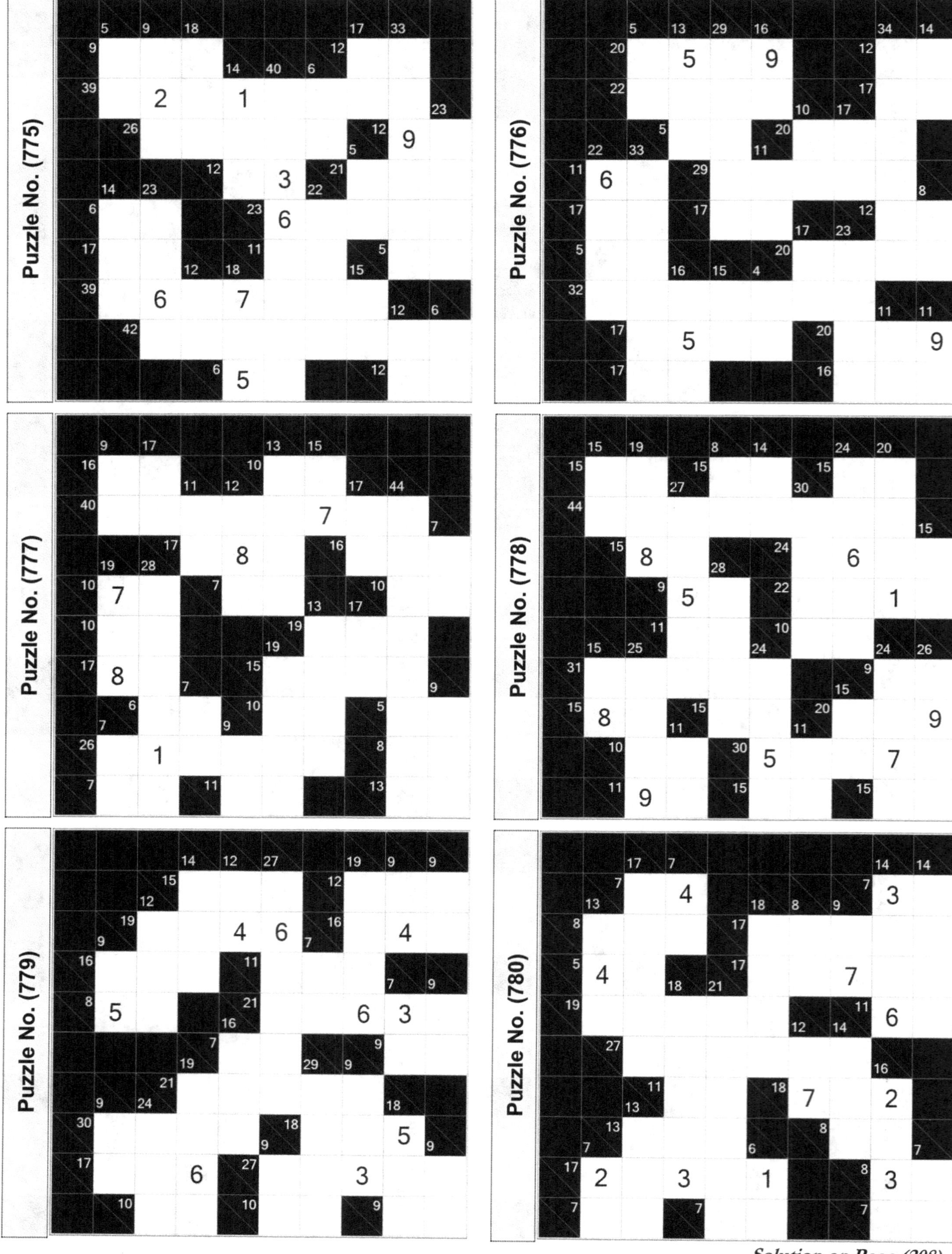

Puzzle No. (775)

Puzzle No. (776)

Puzzle No. (777)

Puzzle No. (778)

Puzzle No. (779)

Puzzle No. (780)

Solution on Page (208)

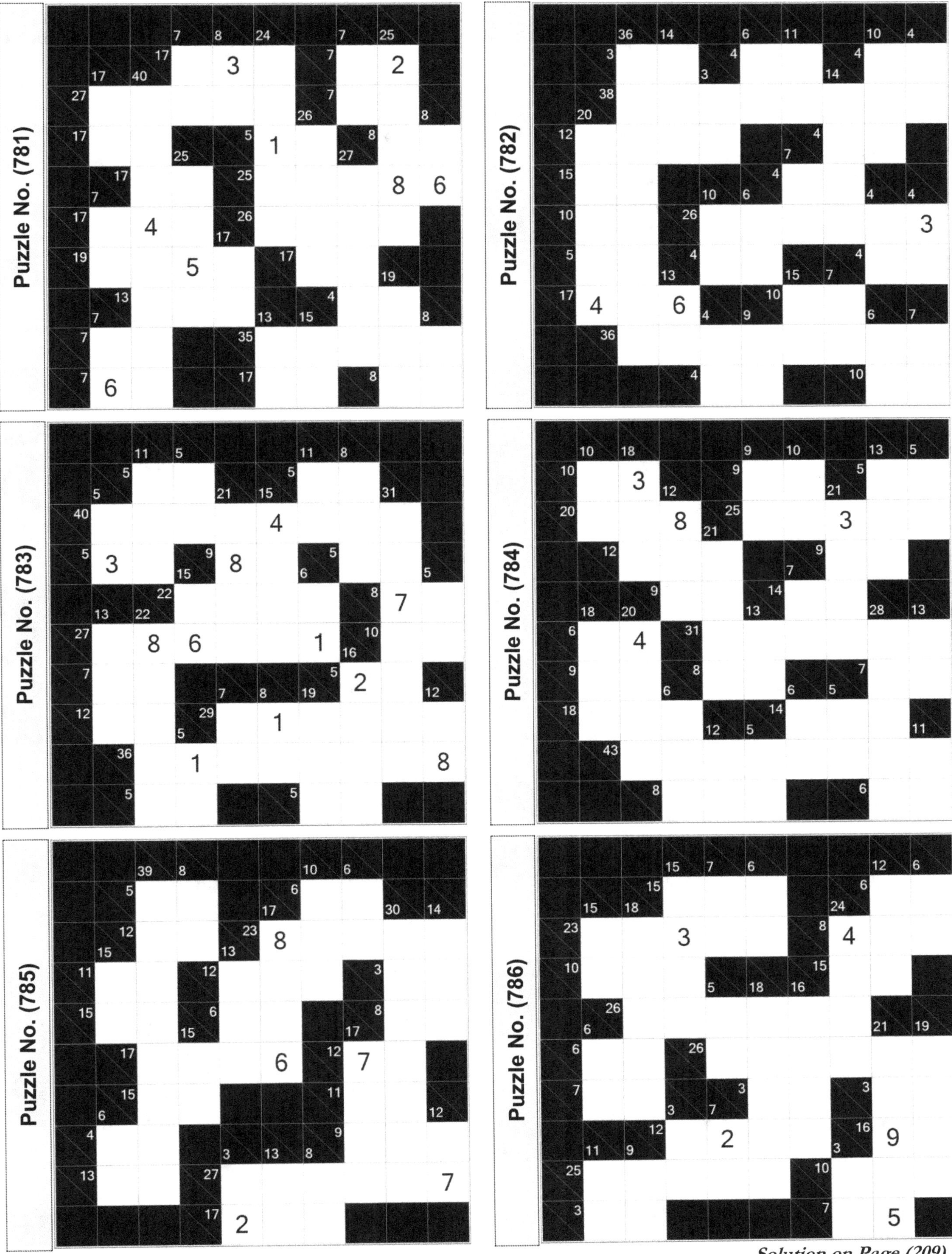

Solution on Page (209)

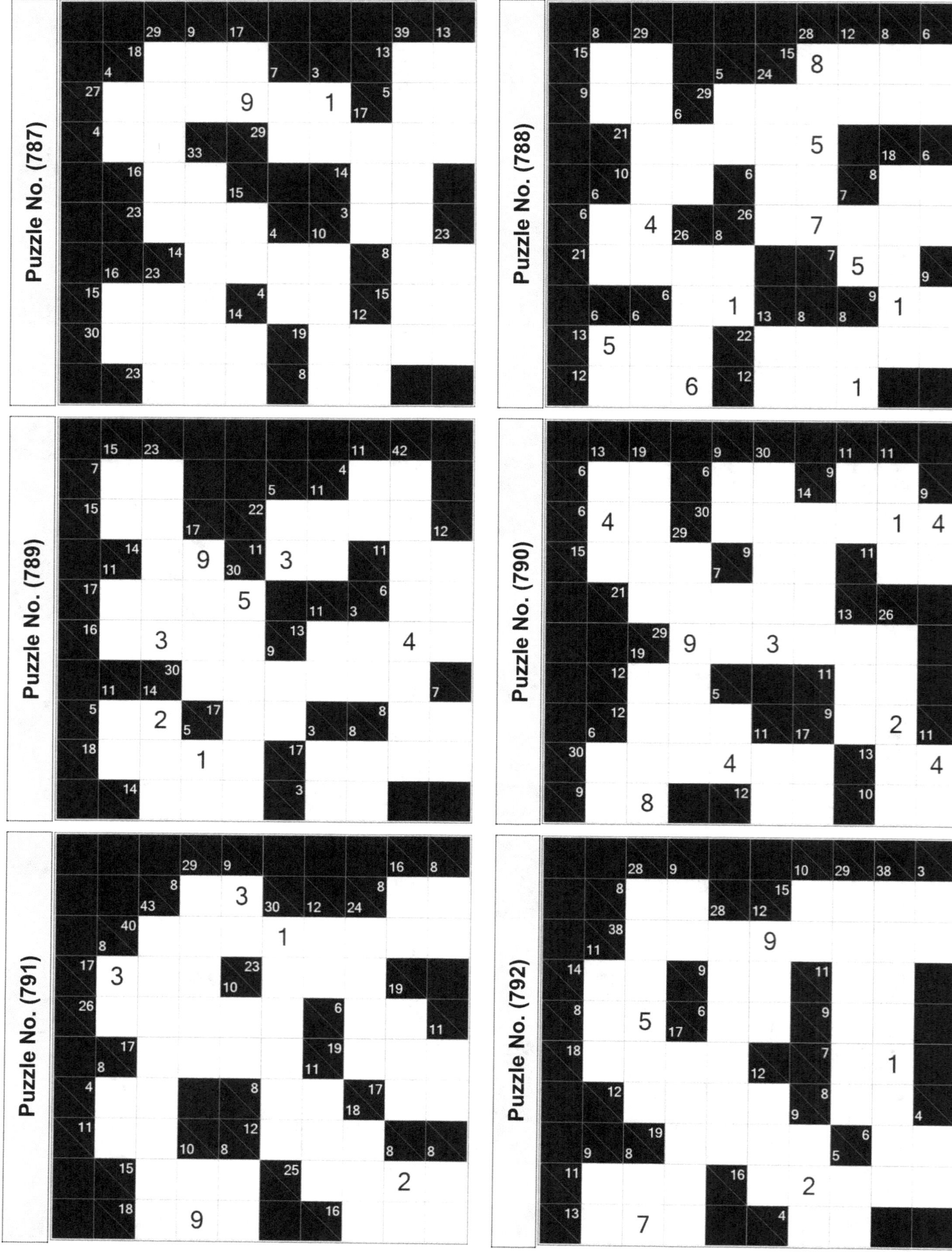

Solution on Page (209)

(134)

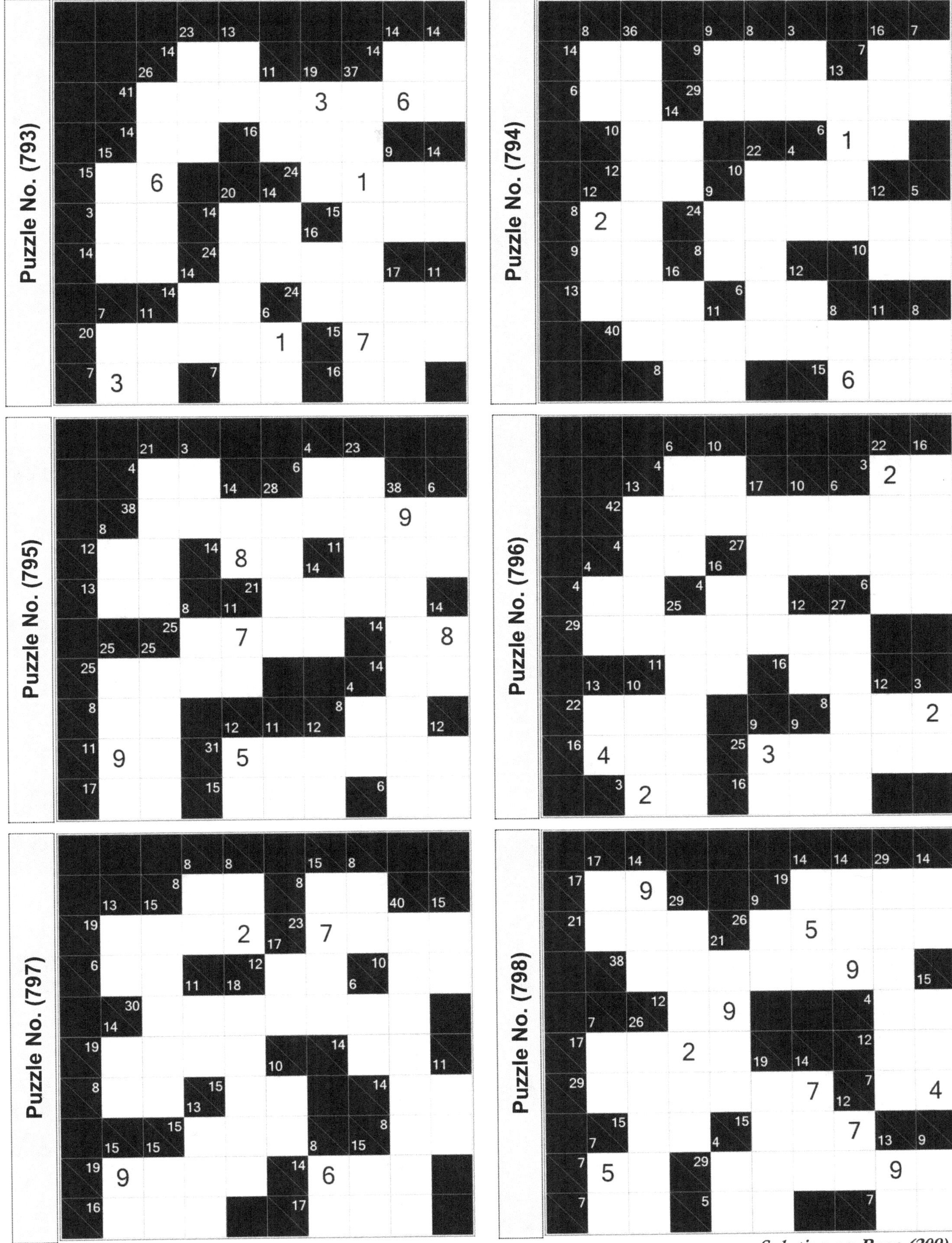

Solution on Page (209)

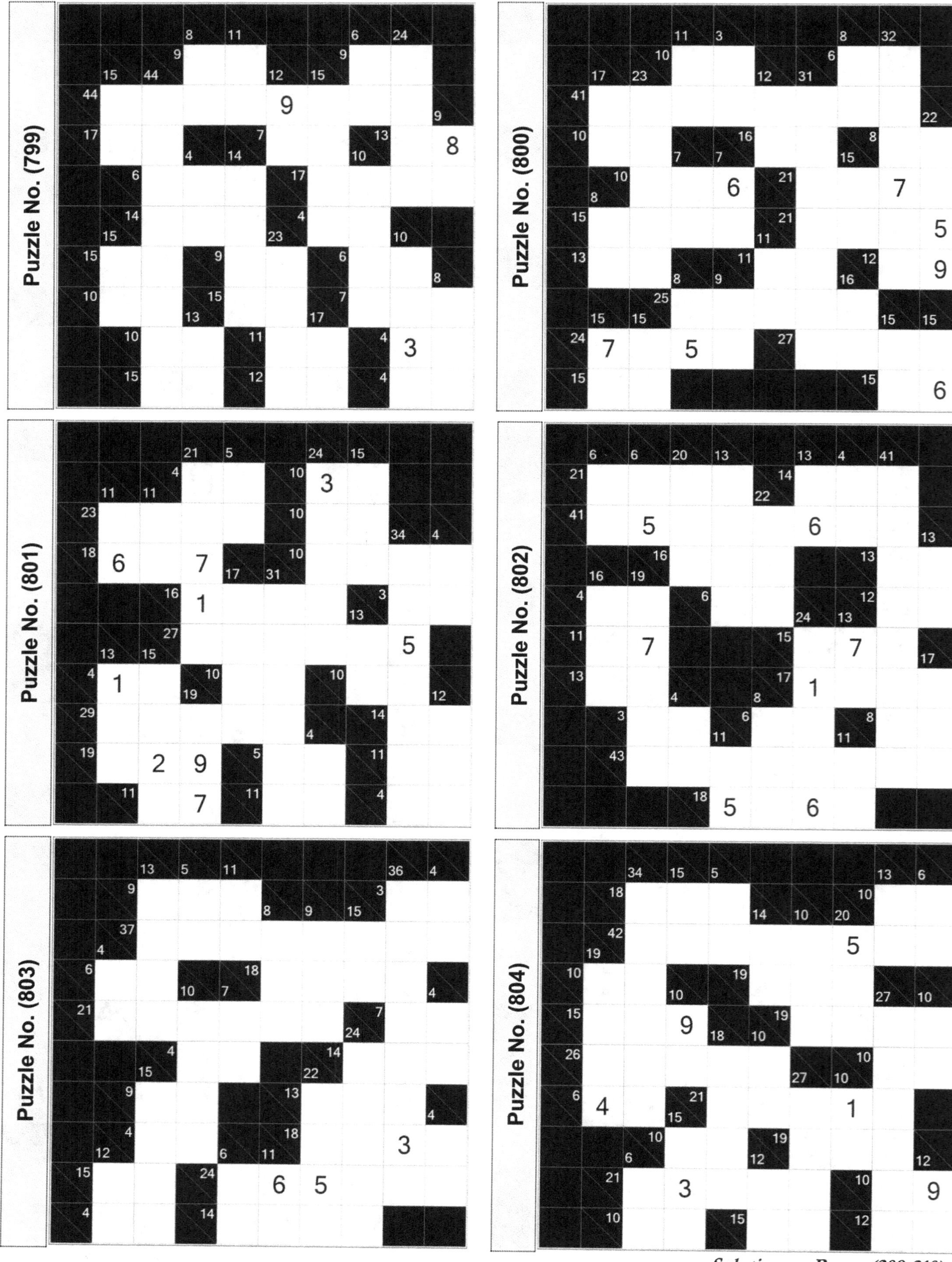

Solution on Pages (209-210)

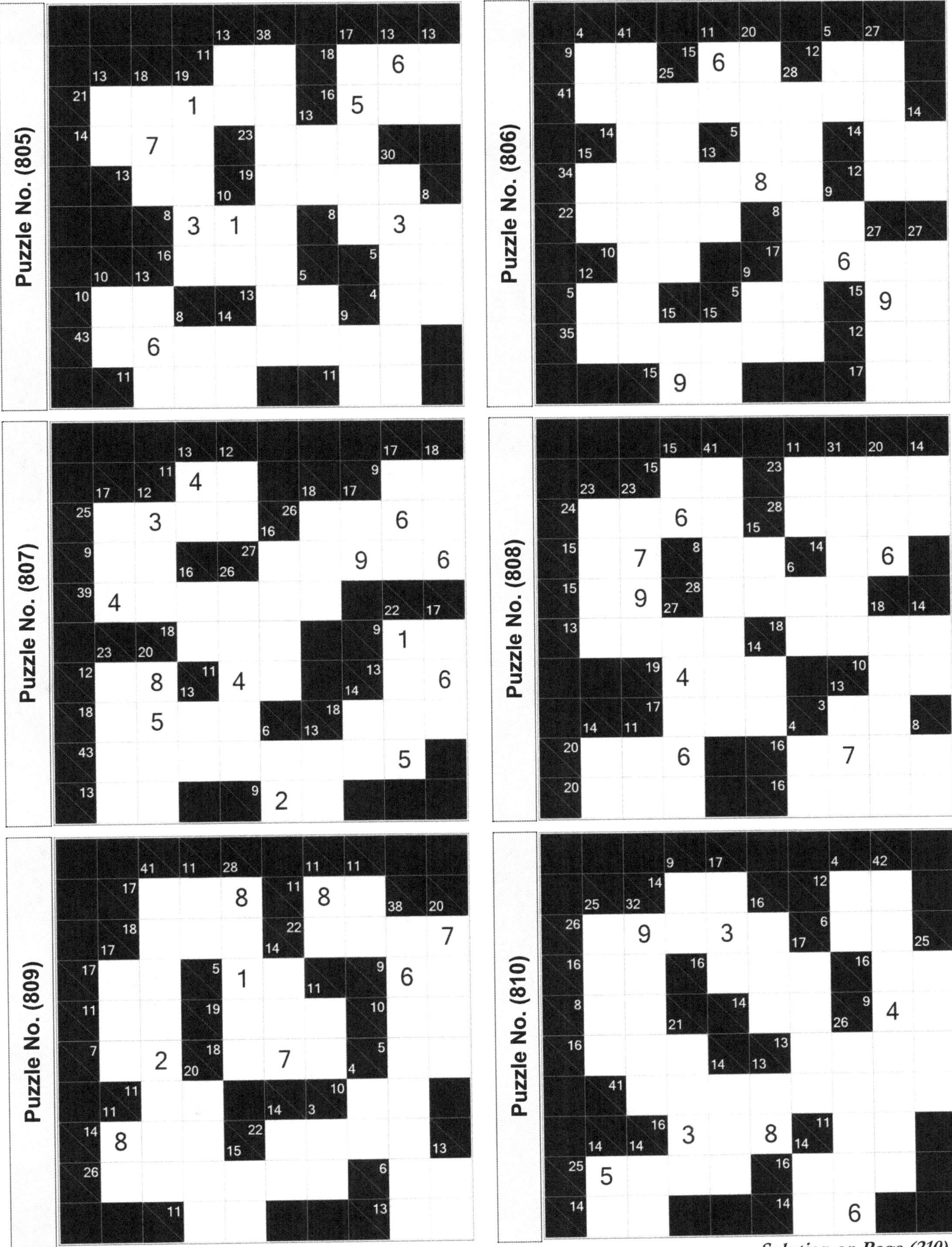

(137)

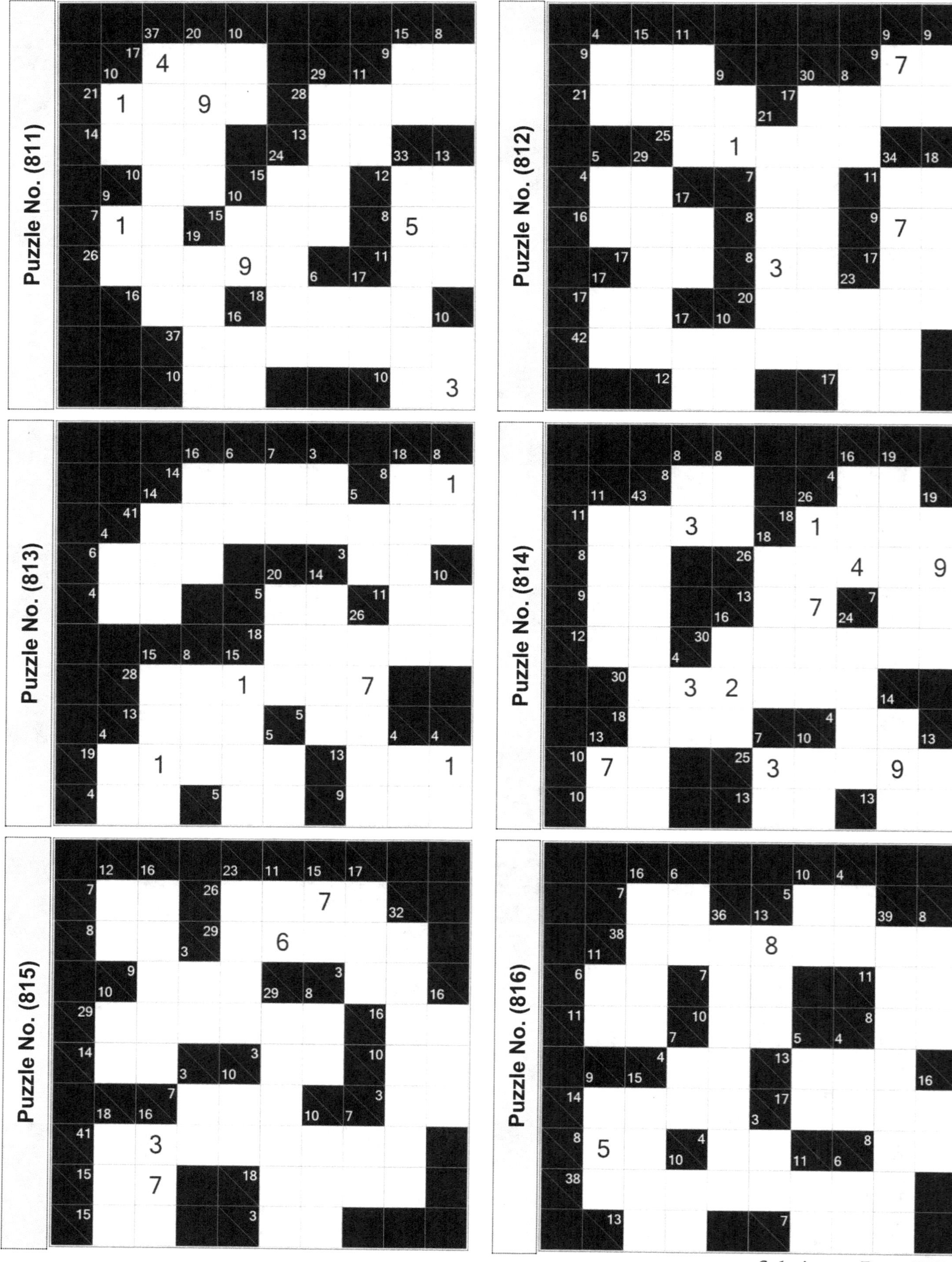

Puzzle No. (811)

Puzzle No. (812)

Puzzle No. (813)

Puzzle No. (814)

Puzzle No. (815)

Puzzle No. (816)

Solution on Page (210)

(138)

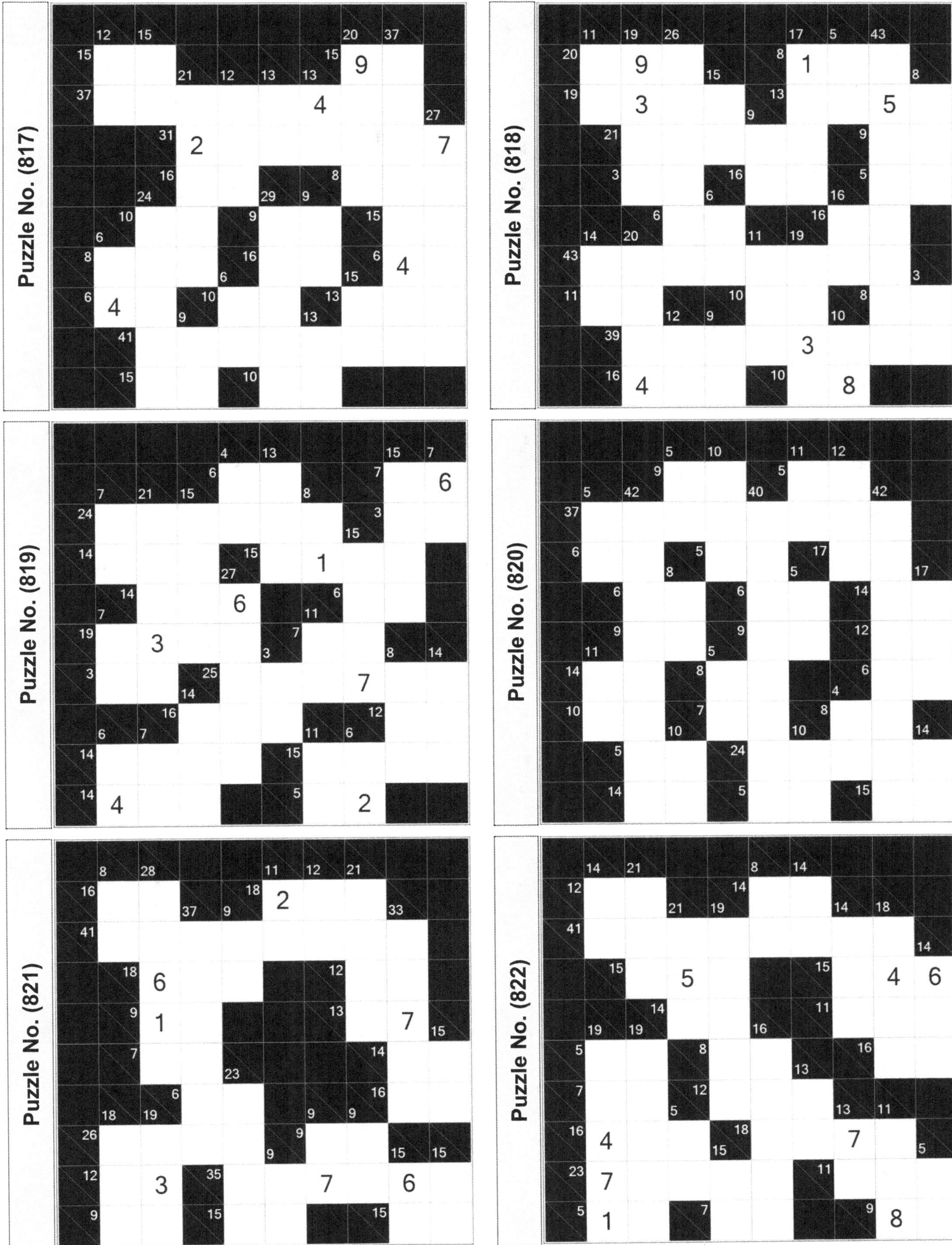

Puzzle No. (817)
Puzzle No. (818)
Puzzle No. (819)
Puzzle No. (820)
Puzzle No. (821)
Puzzle No. (822)
Solution on Pages (210-211)

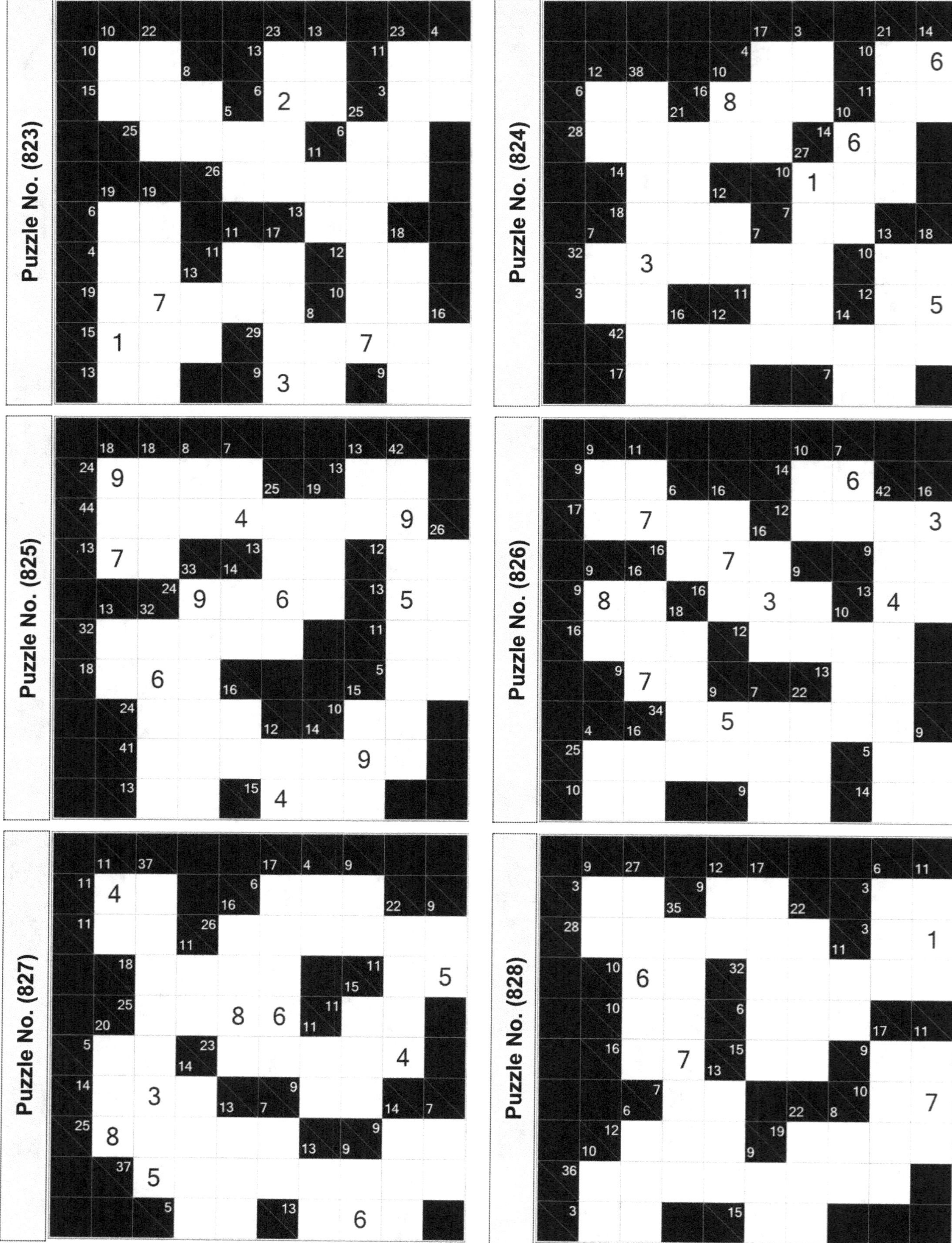

Puzzle No. (823)

Puzzle No. (824)

Puzzle No. (825)

Puzzle No. (826)

Puzzle No. (827)

Puzzle No. (828)

Solution on Page (211)

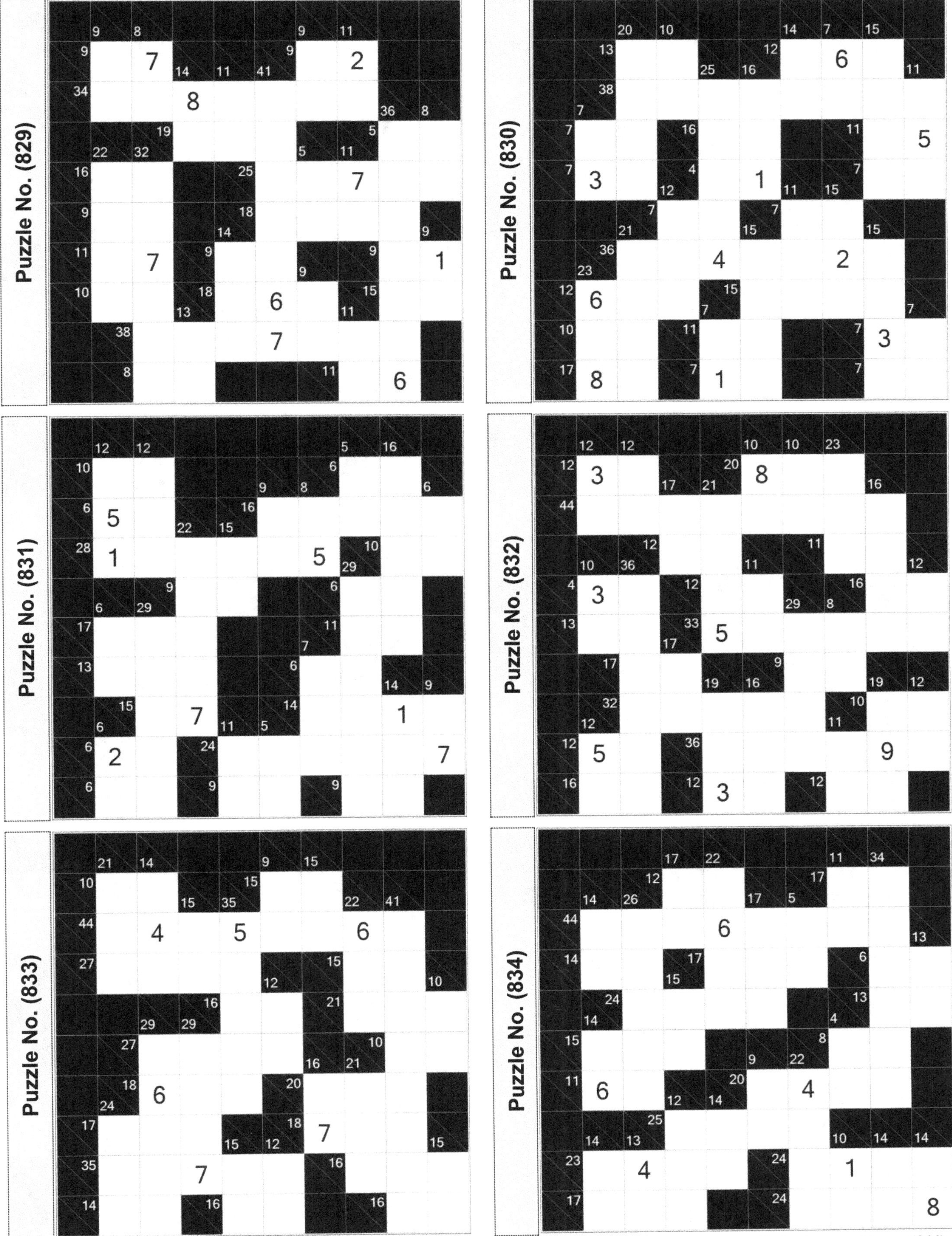

Puzzle No. (829)

Puzzle No. (830)

Puzzle No. (831)

Puzzle No. (832)

Puzzle No. (833)

Puzzle No. (834)

Solution on Page (211)

(141)

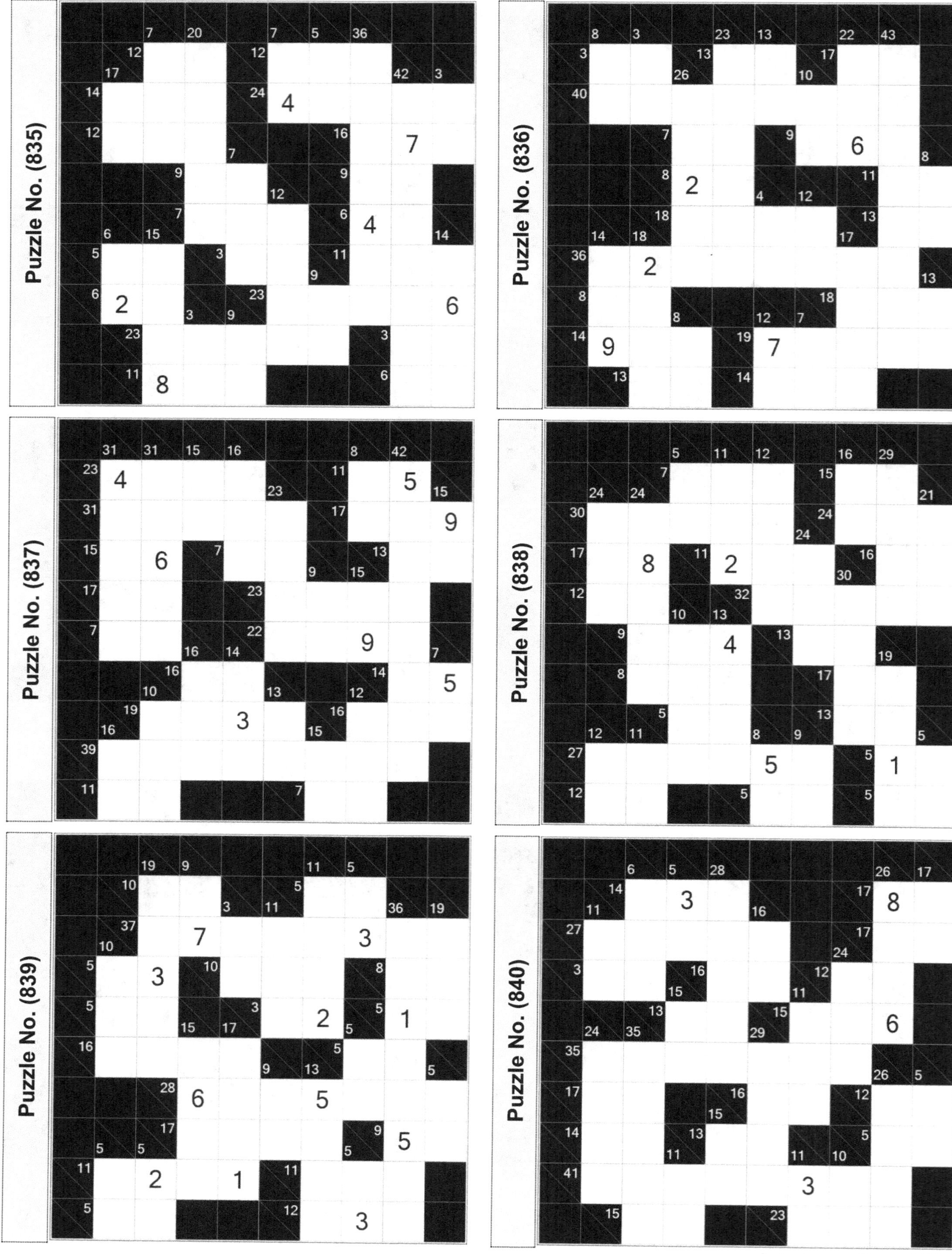

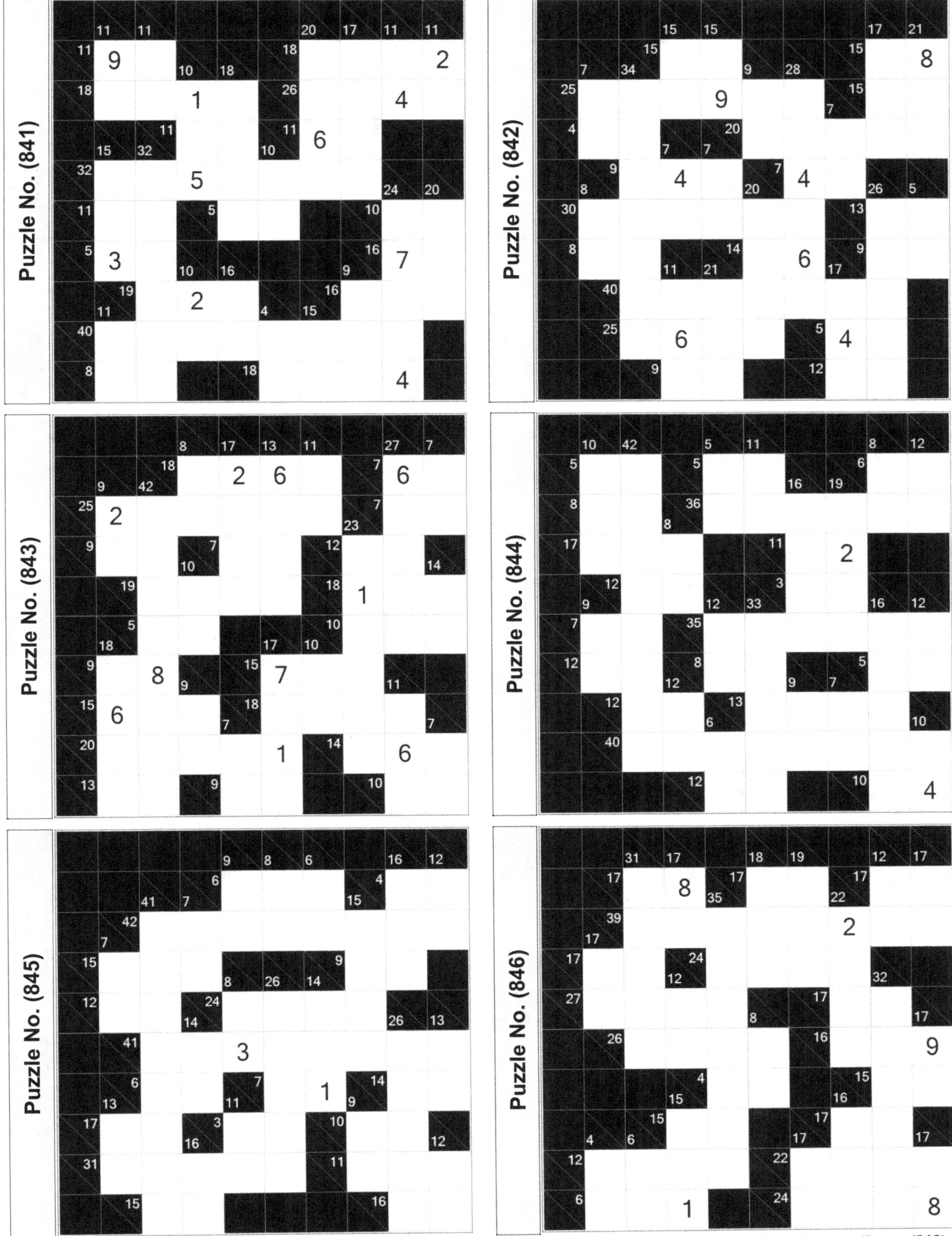

Puzzle No. (841)

Puzzle No. (842)

Puzzle No. (843)

Puzzle No. (844)

Puzzle No. (845)

Puzzle No. (846)

Solution on Page (212)

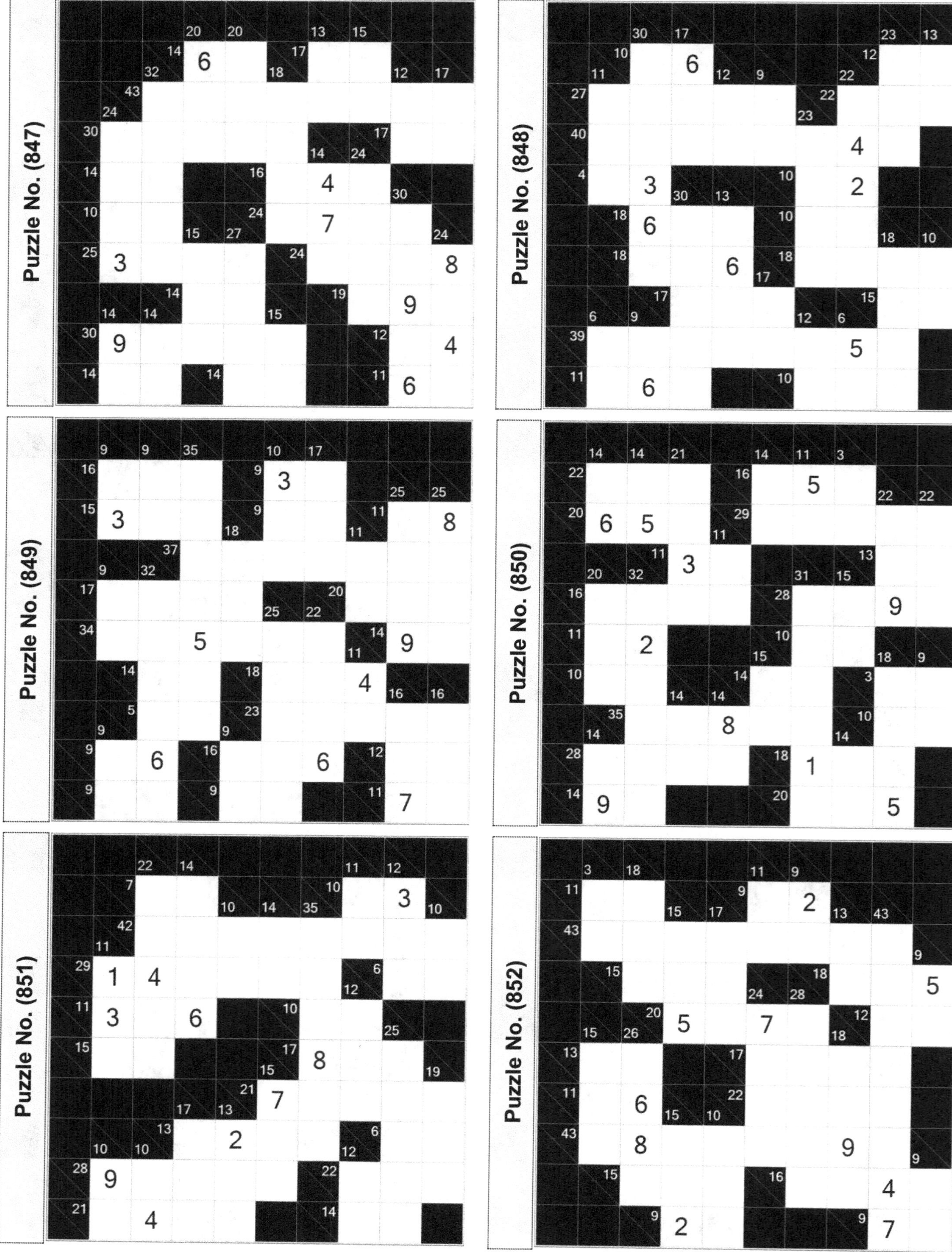

Puzzle No. (847)

Puzzle No. (848)

Puzzle No. (849)

Puzzle No. (850)

Puzzle No. (851)

Puzzle No. (852)

Solution on Page (212)

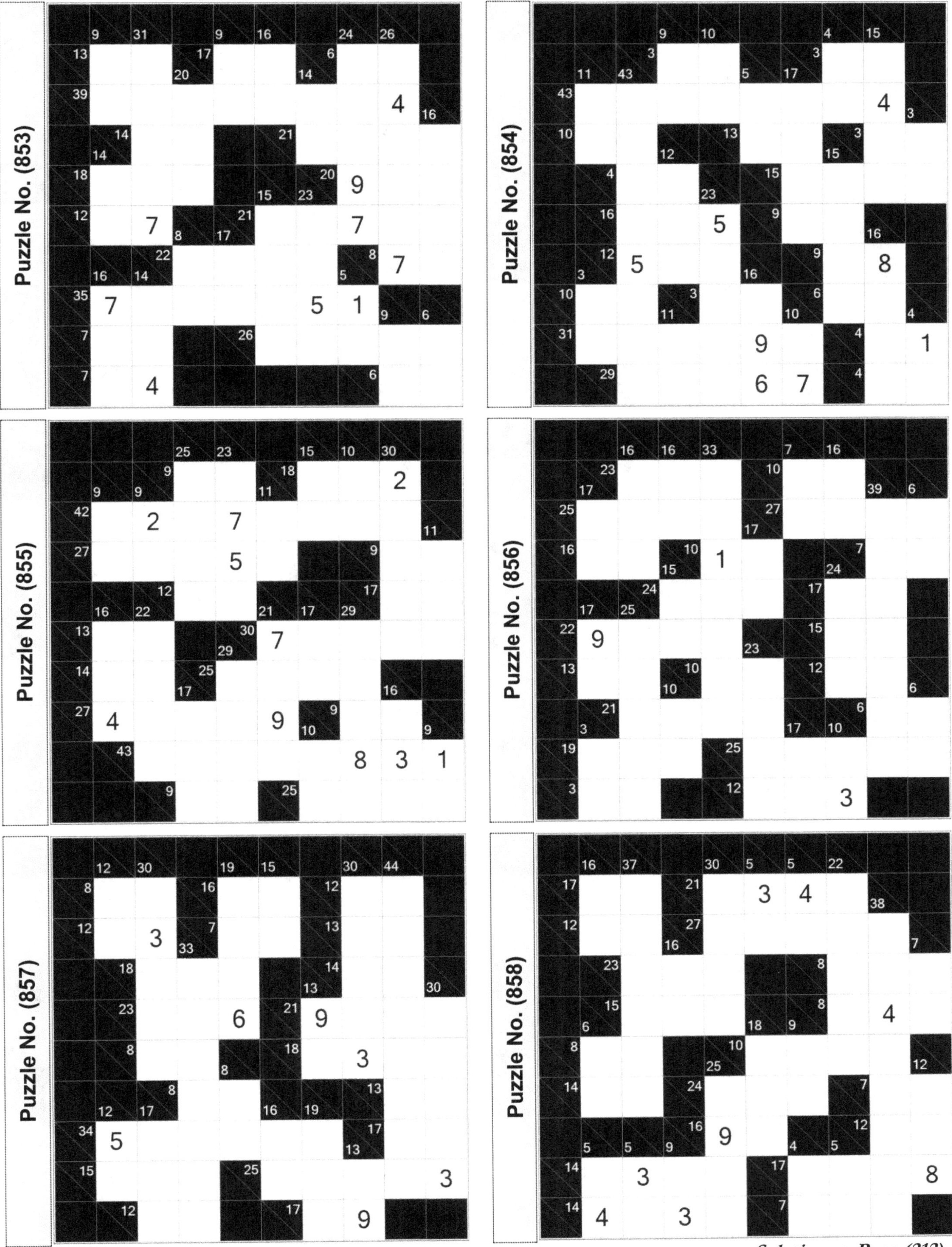

Puzzle No. (853)

Puzzle No. (854)

Puzzle No. (855)

Puzzle No. (856)

Puzzle No. (857)

Puzzle No. (858)

Solution on Page (212)

(145)

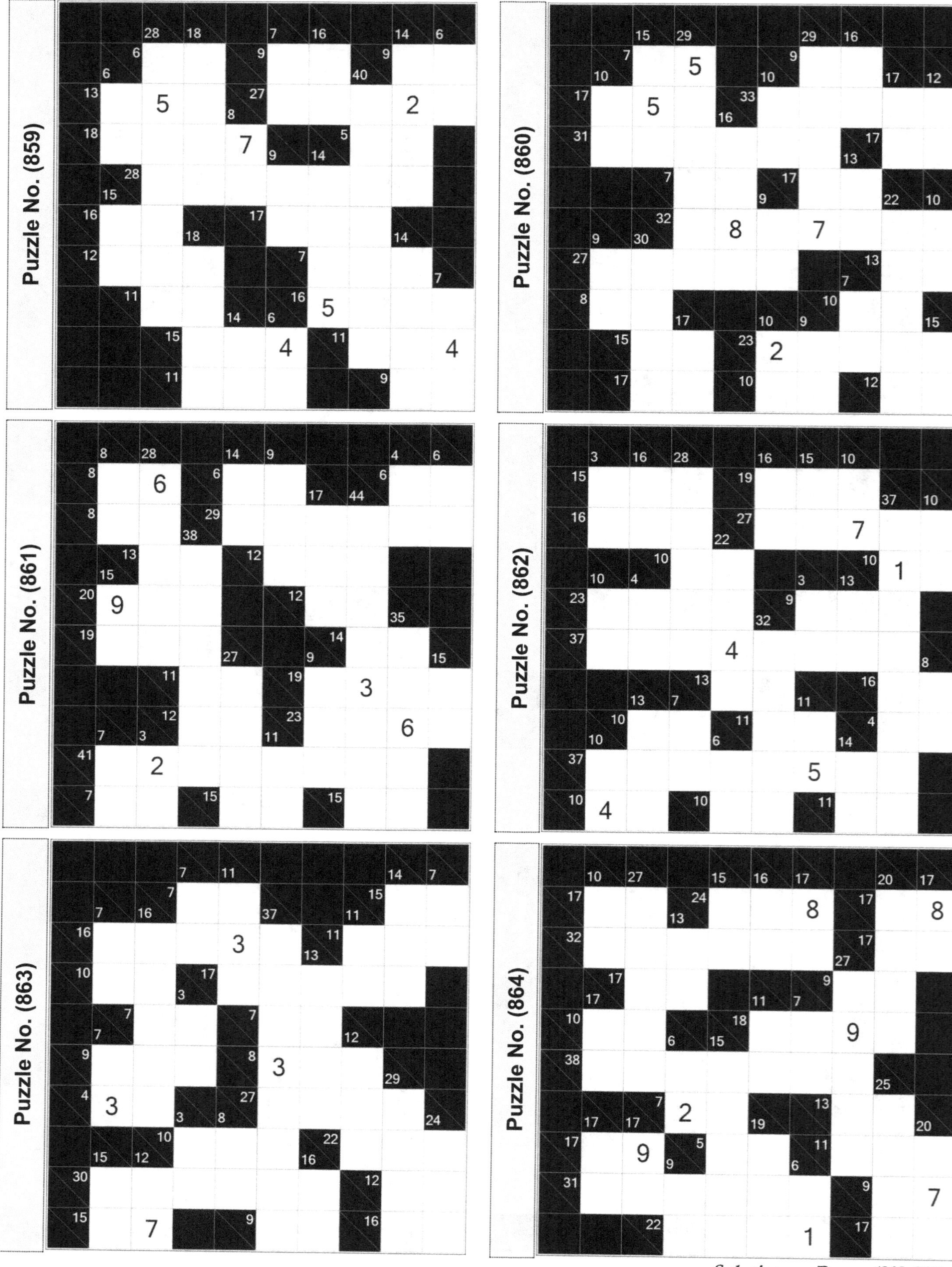

Puzzle No. (859)

Puzzle No. (860)

Puzzle No. (861)

Puzzle No. (862)

Puzzle No. (863)

Puzzle No. (864)

Solution on Pages (212-213)

(146)

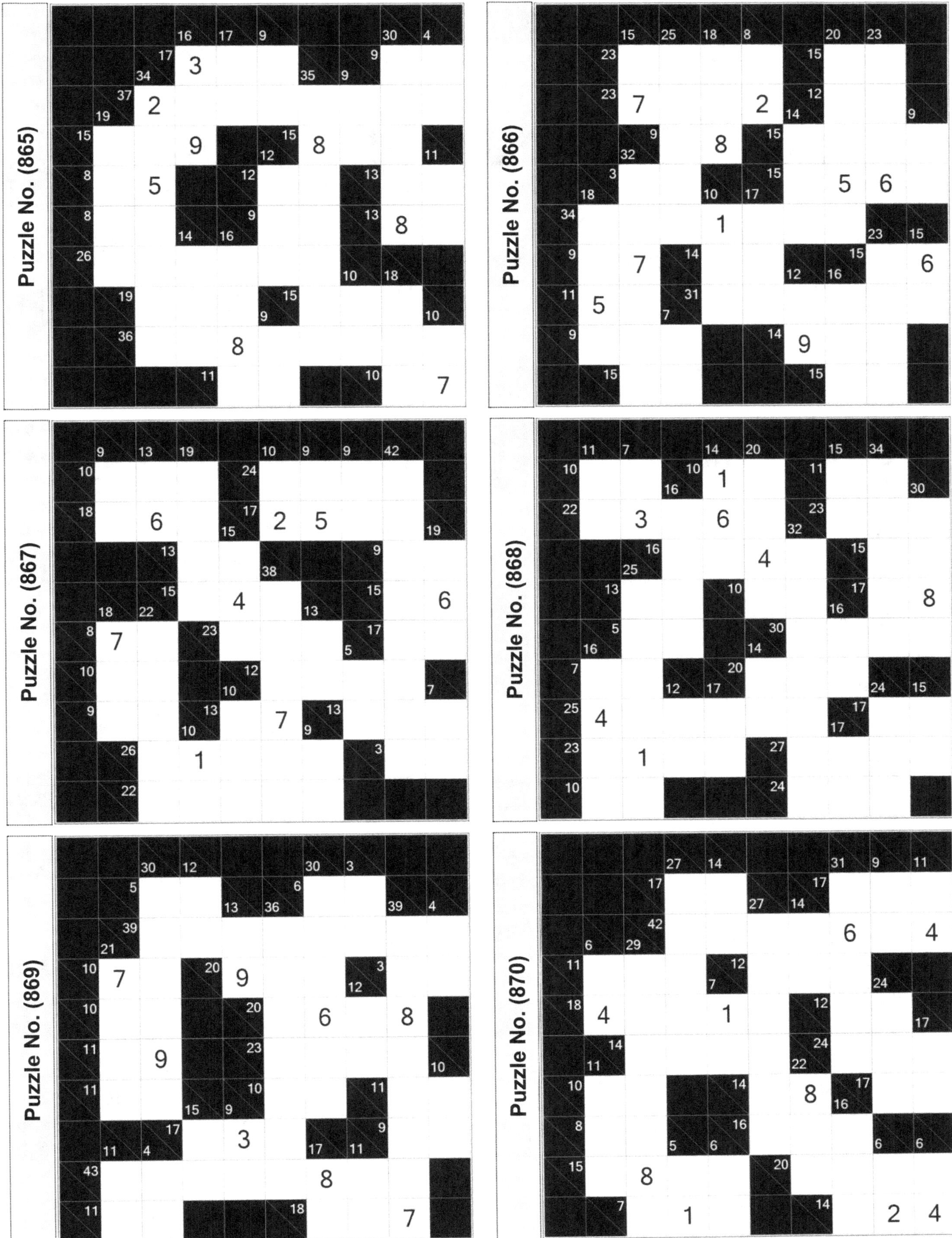

Solution on Page (213)

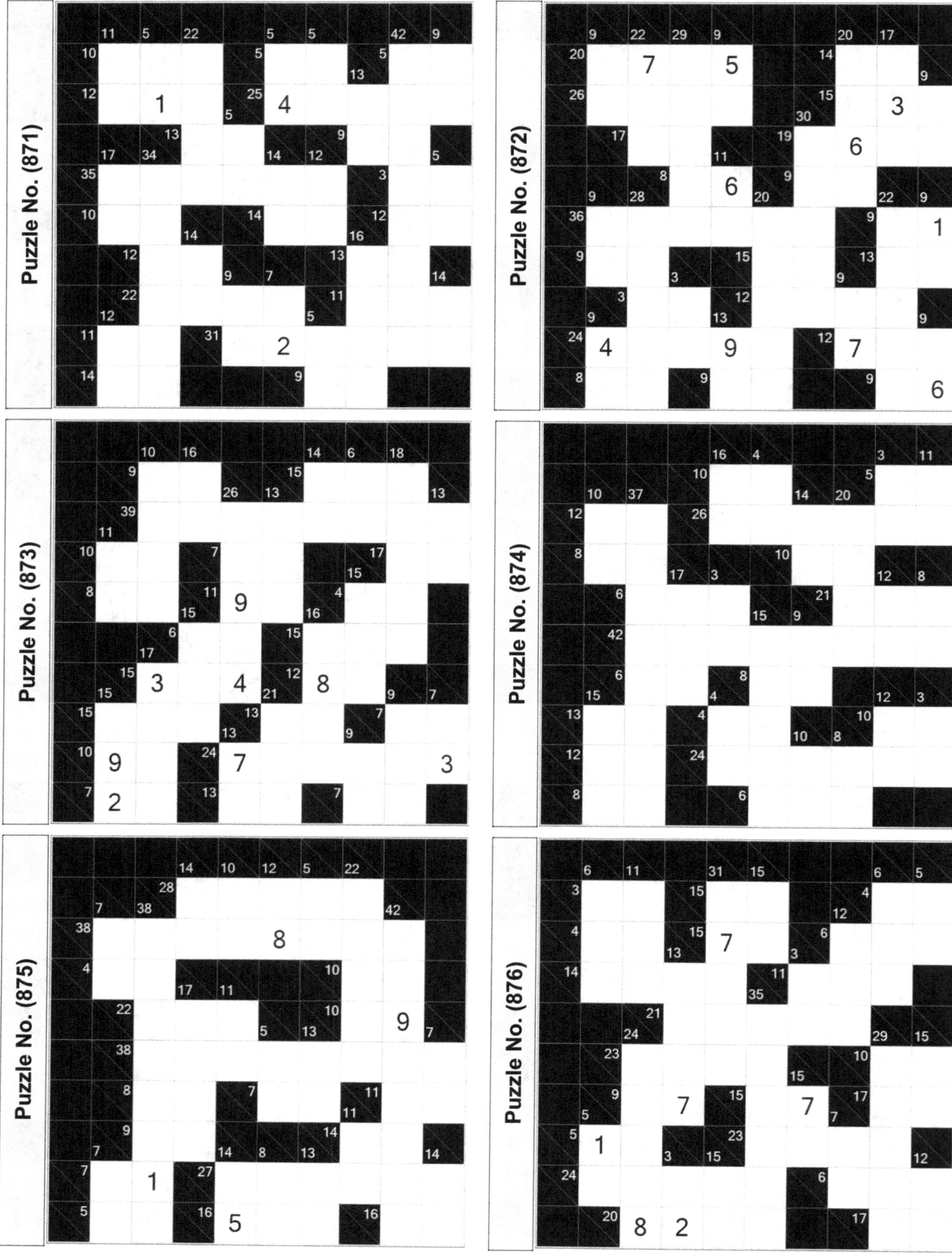

Puzzle No. (871)

Puzzle No. (872)

Puzzle No. (873)

Puzzle No. (874)

Puzzle No. (875)

Puzzle No. (876)

Solution on Page (213)

(148)

Solution on Pages (213-214)

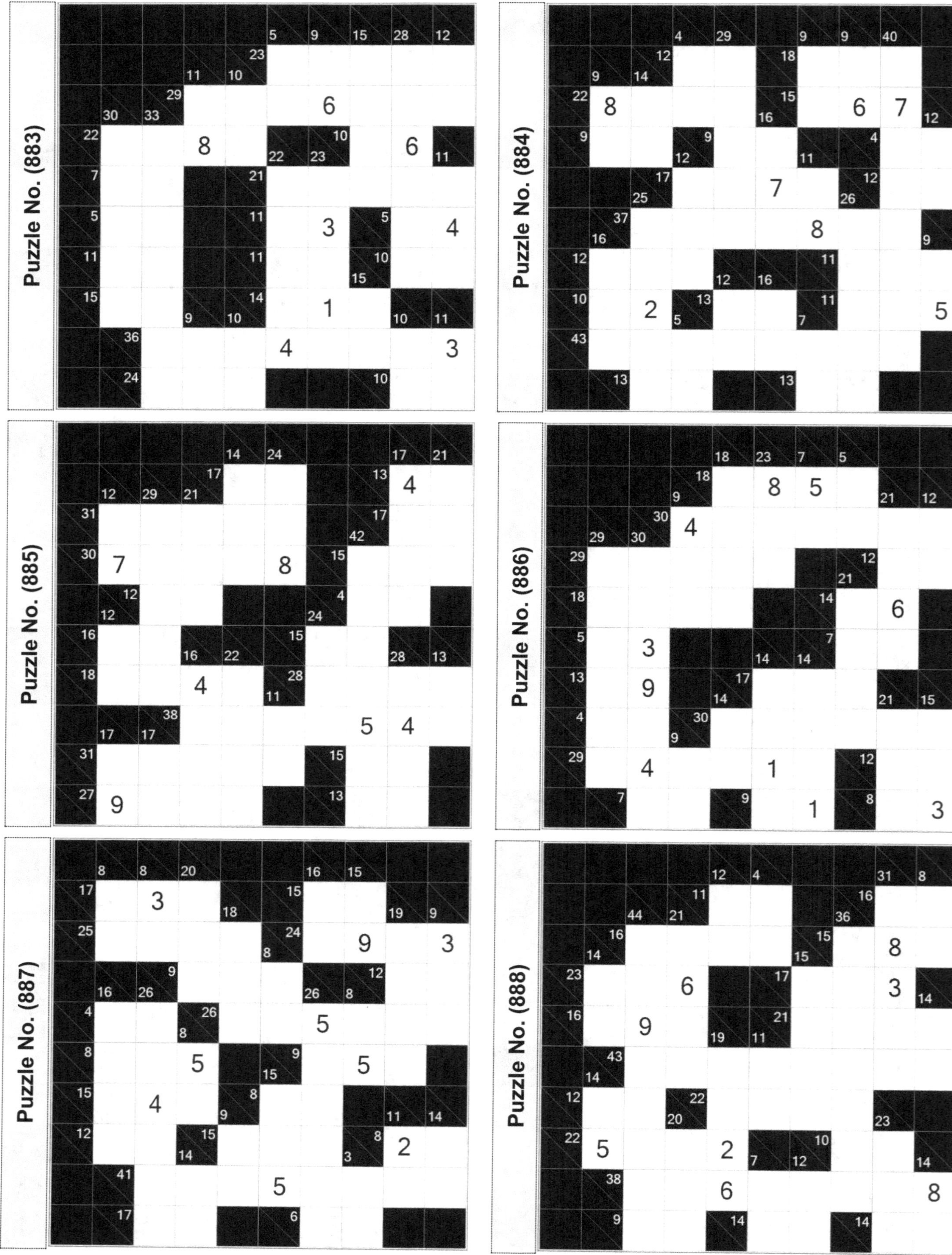

Solution on Page (214)

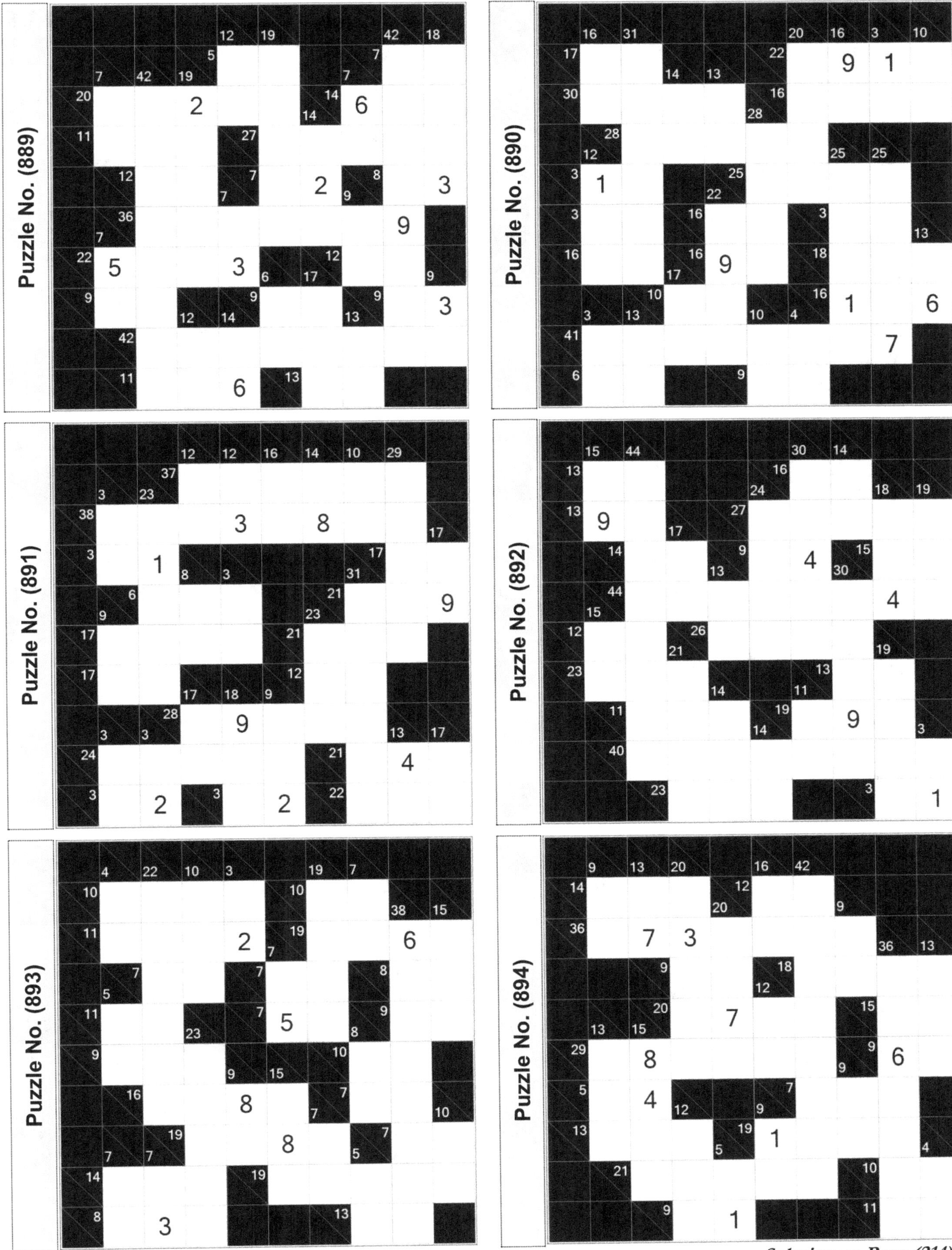

Puzzle No. (889)

Puzzle No. (890)

Puzzle No. (891)

Puzzle No. (892)

Puzzle No. (893)

Puzzle No. (894)

Solution on Page (214)

(151)

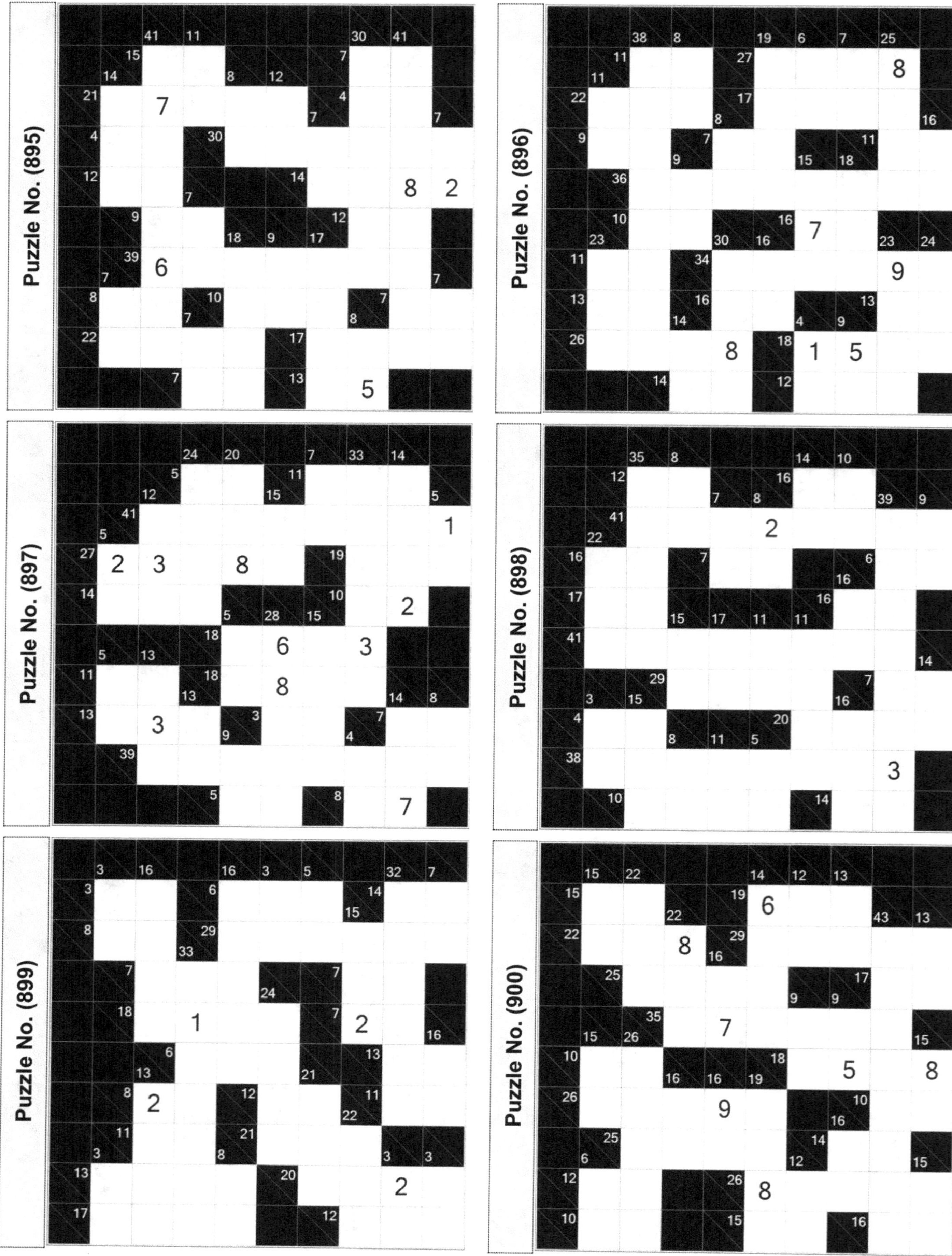

Puzzle No. (895)
Puzzle No. (896)
Puzzle No. (897)
Puzzle No. (898)
Puzzle No. (899)
Puzzle No. (900)

Solution on Page (215)

(153)

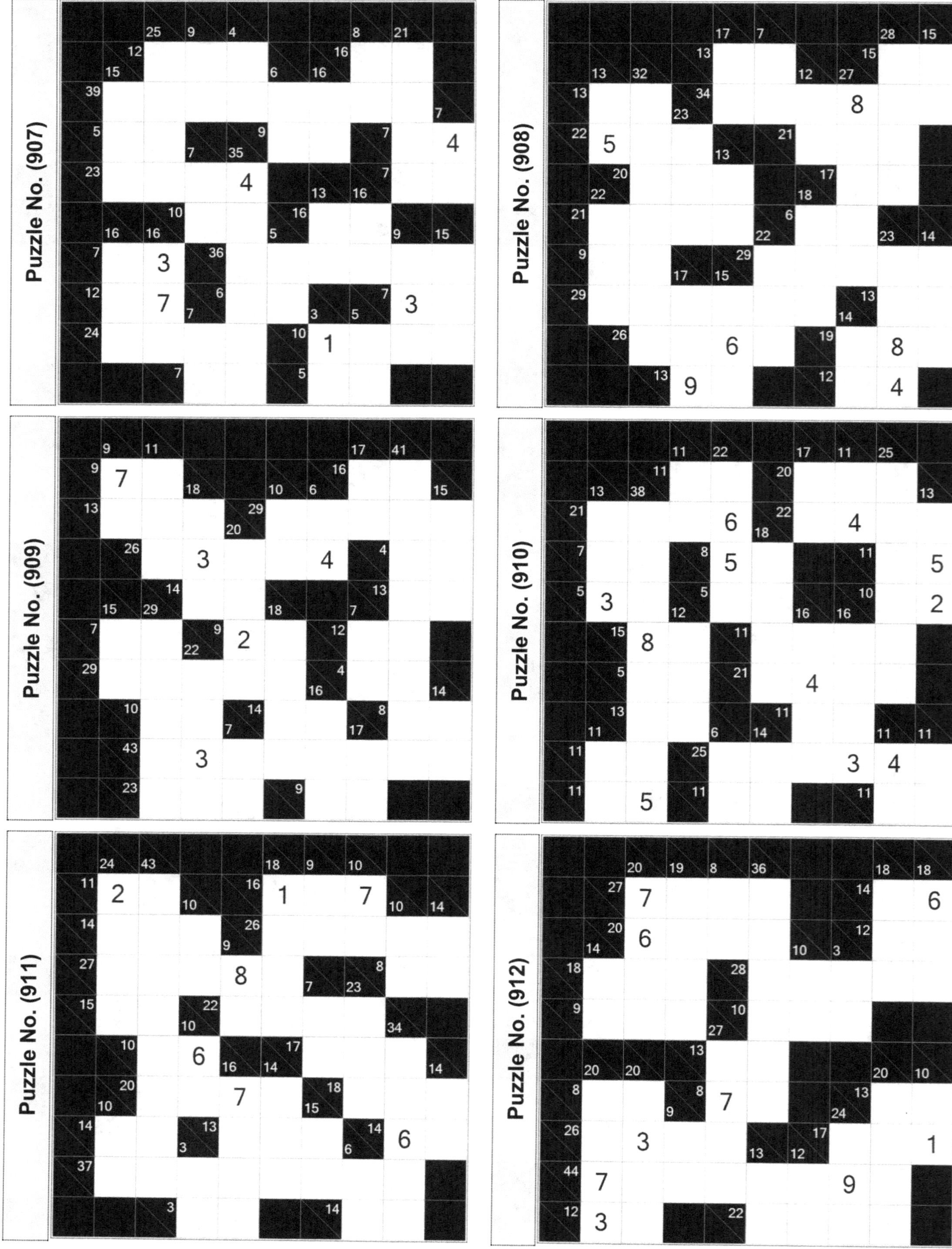

Puzzle No. (907)
Puzzle No. (908)
Puzzle No. (909)
Puzzle No. (910)
Puzzle No. (911)
Puzzle No. (912)
Solution on Page (215)

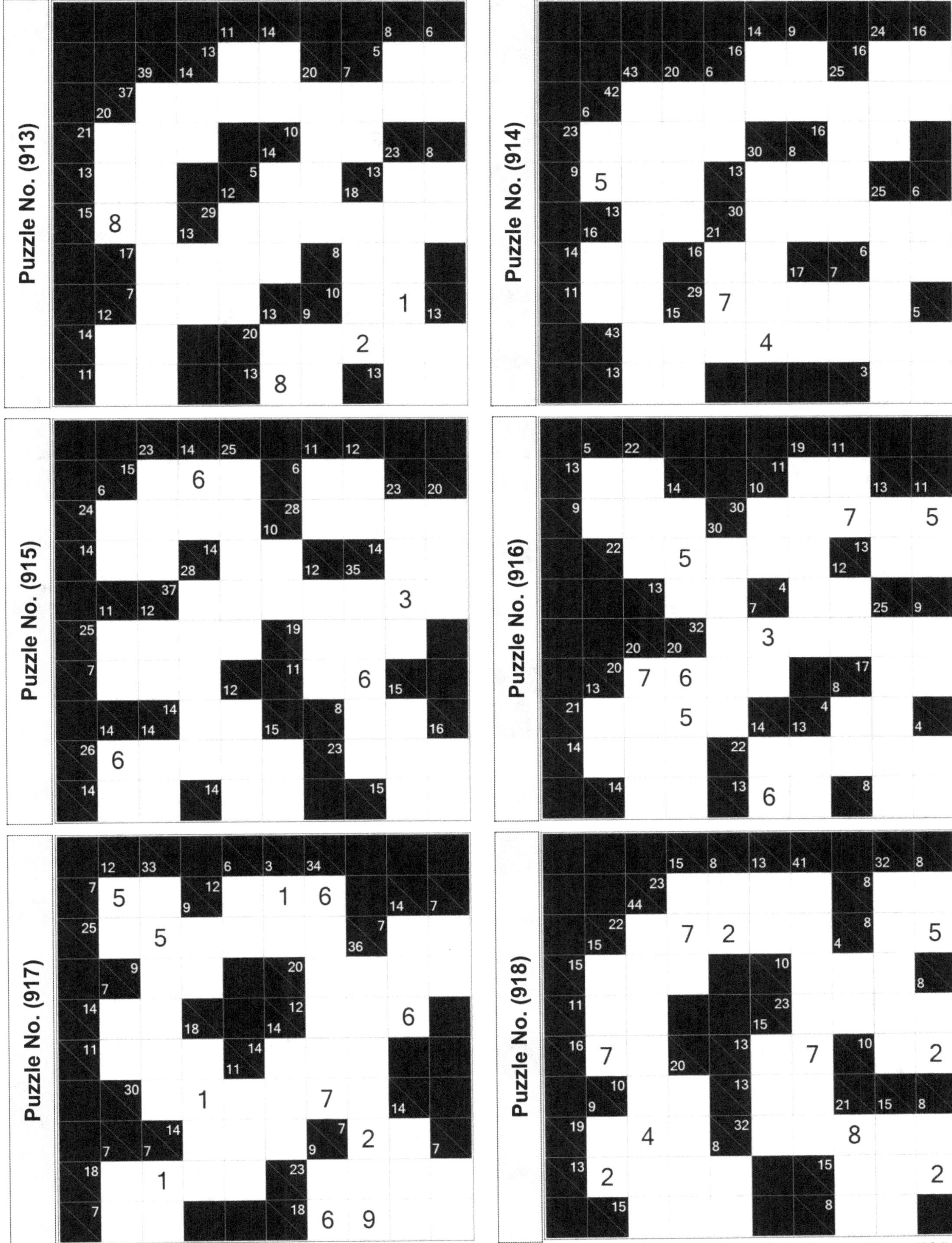

Puzzle No. (913)

Puzzle No. (914)

Puzzle No. (915)

Puzzle No. (916)

Puzzle No. (917)

Puzzle No. (918)

Solution on Page (215)

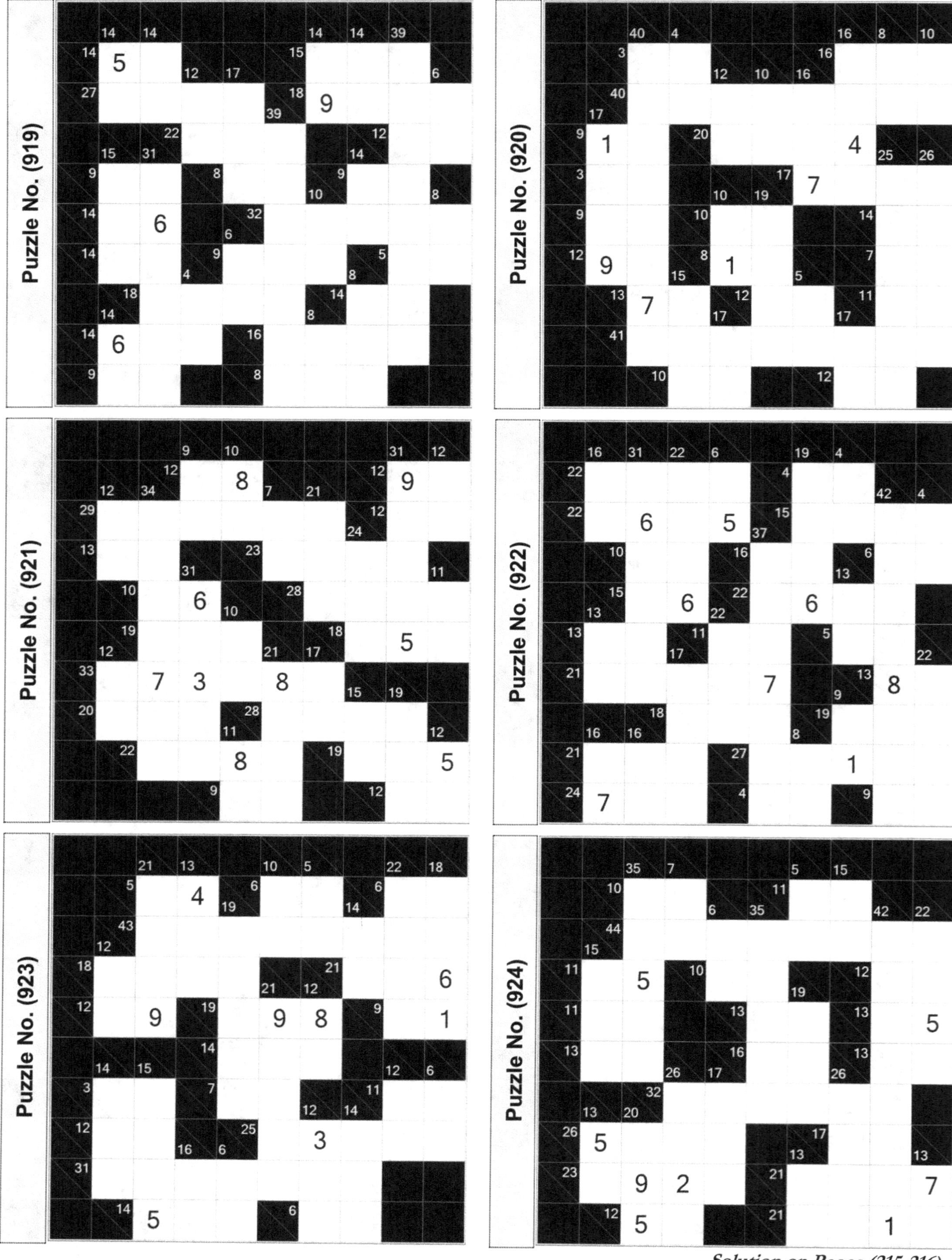

Puzzle No. (919)

Puzzle No. (920)

Puzzle No. (921)

Puzzle No. (922)

Puzzle No. (923)

Puzzle No. (924)

Solution on Pages (215-216)

(156)

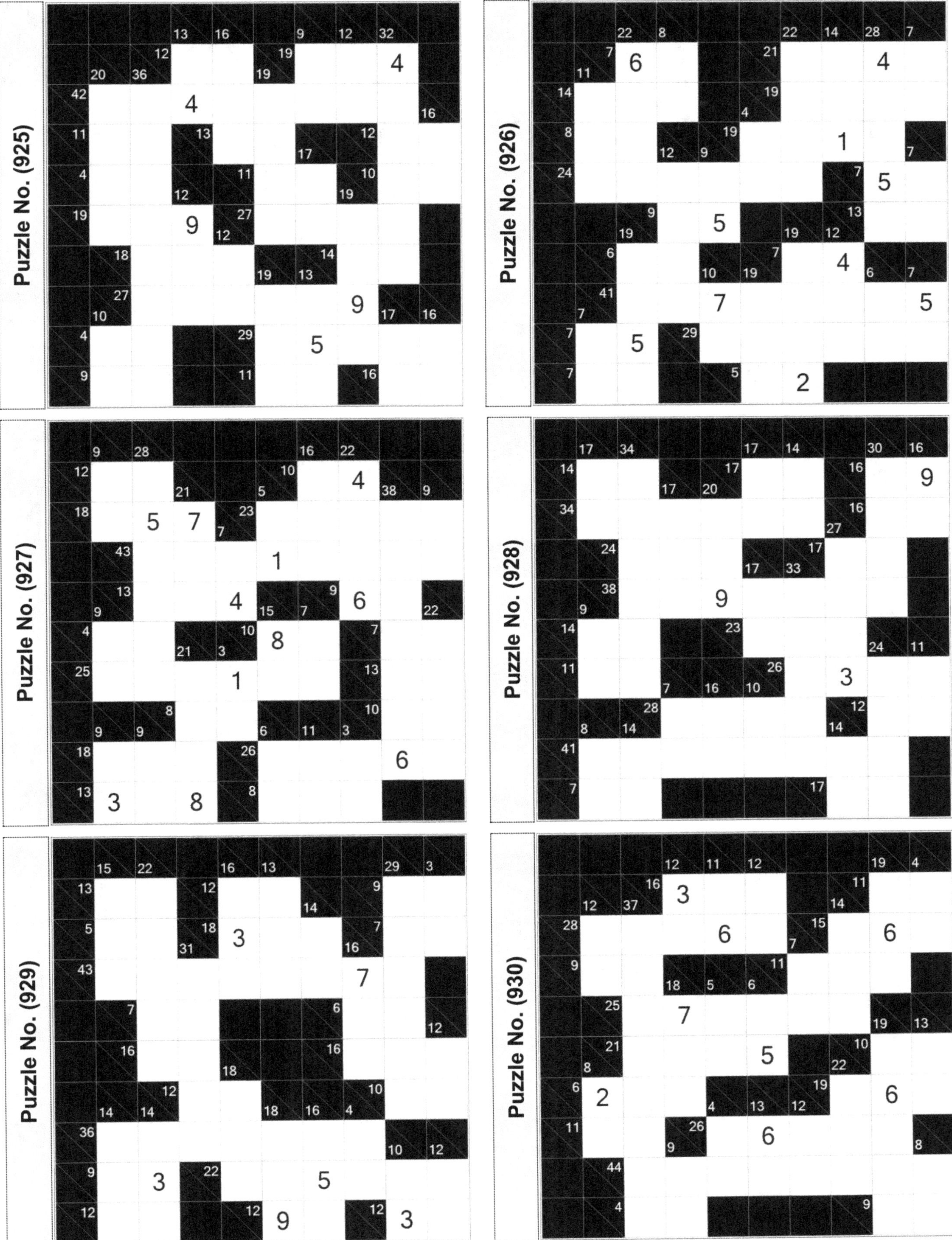

Solution on Page (216)

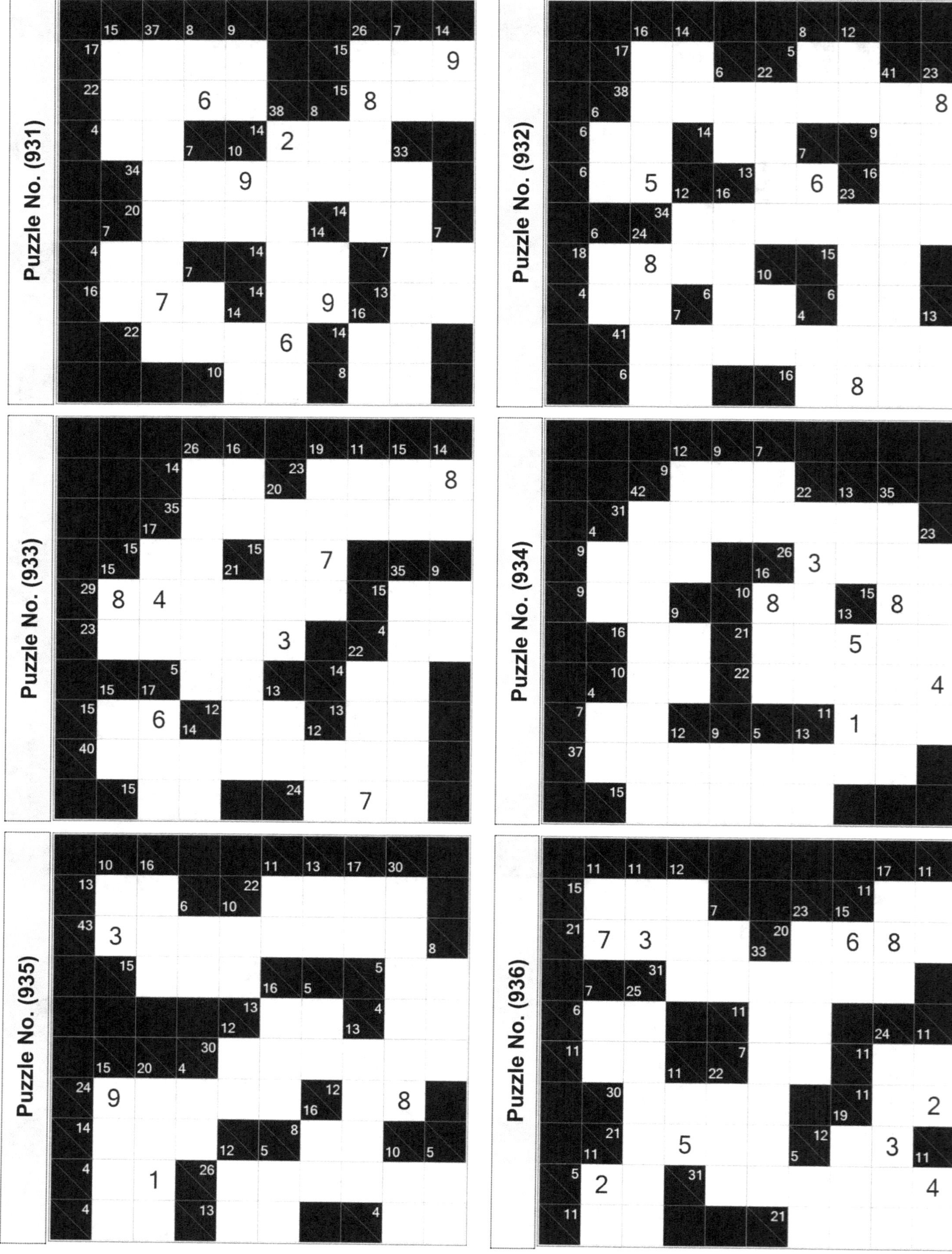

Solution on Page (216)

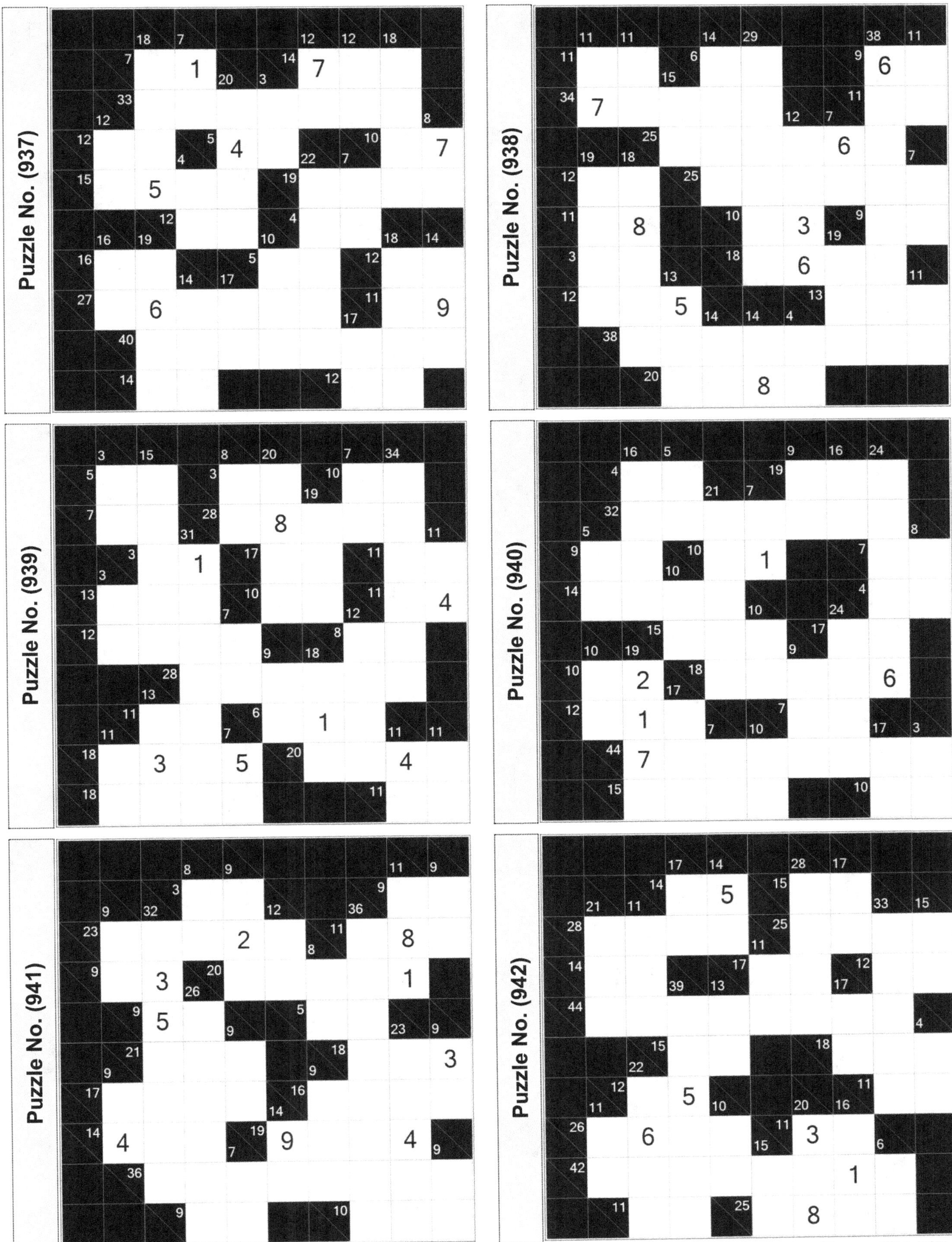

Puzzle No. (937)

Puzzle No. (938)

Puzzle No. (939)

Puzzle No. (940)

Puzzle No. (941)

Puzzle No. (942)

Solution on Pages (216-217)

Puzzle No. (943)

Puzzle No. (944)

Puzzle No. (945)

Puzzle No. (946)

Puzzle No. (947)

Puzzle No. (948)

Solution on Page (217)

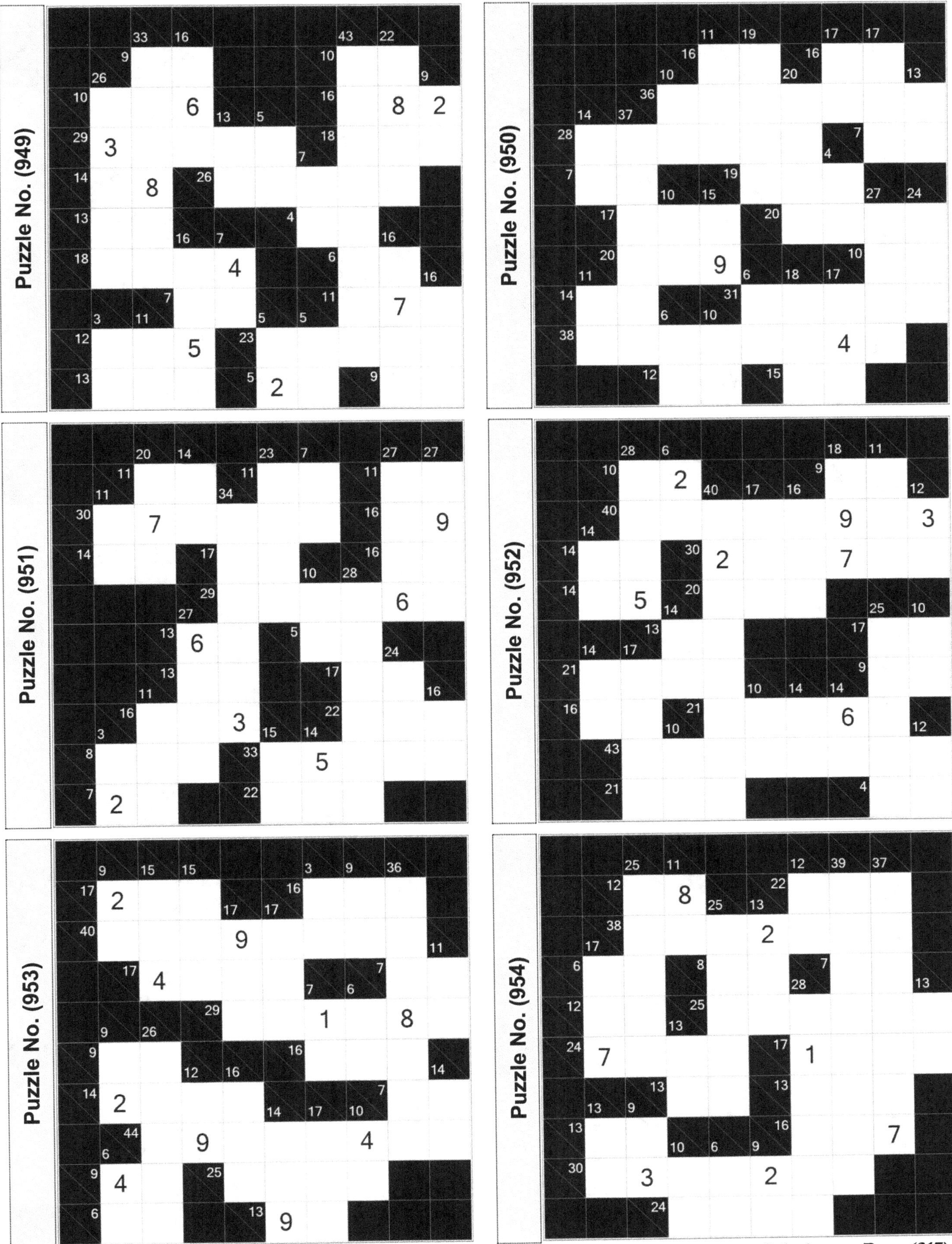

Puzzle No. (949)
Puzzle No. (950)
Puzzle No. (951)
Puzzle No. (952)
Puzzle No. (953)
Puzzle No. (954)
Solution on Page (217)

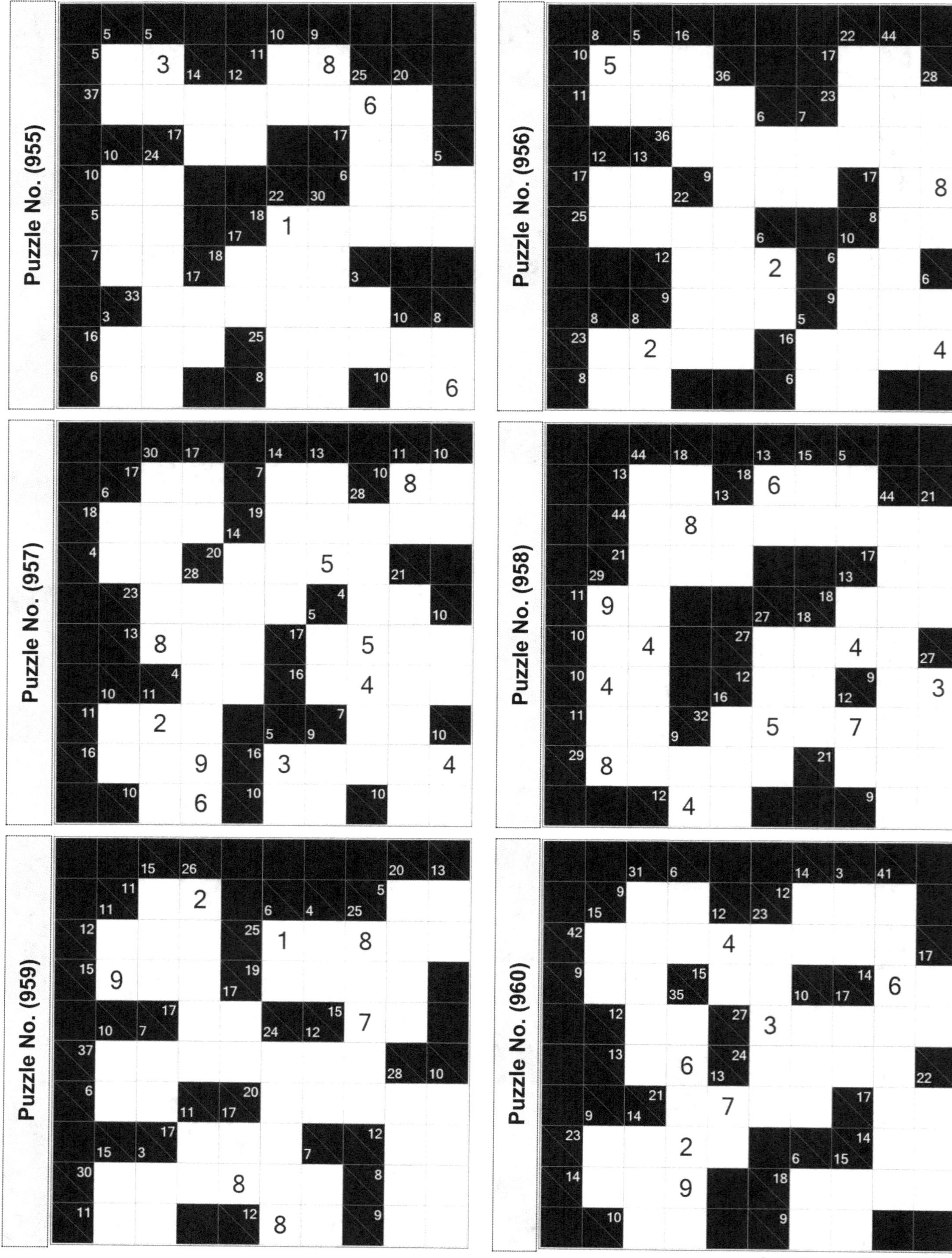

Puzzle No. (955)
Puzzle No. (956)
Puzzle No. (957)
Puzzle No. (958)
Puzzle No. (959)
Puzzle No. (960)
Solution on Page (217)

Puzzle No. (961)

Puzzle No. (962)

Puzzle No. (963)

Puzzle No. (964)

Puzzle No. (965)

Puzzle No. (966)

Solution on Page (218)

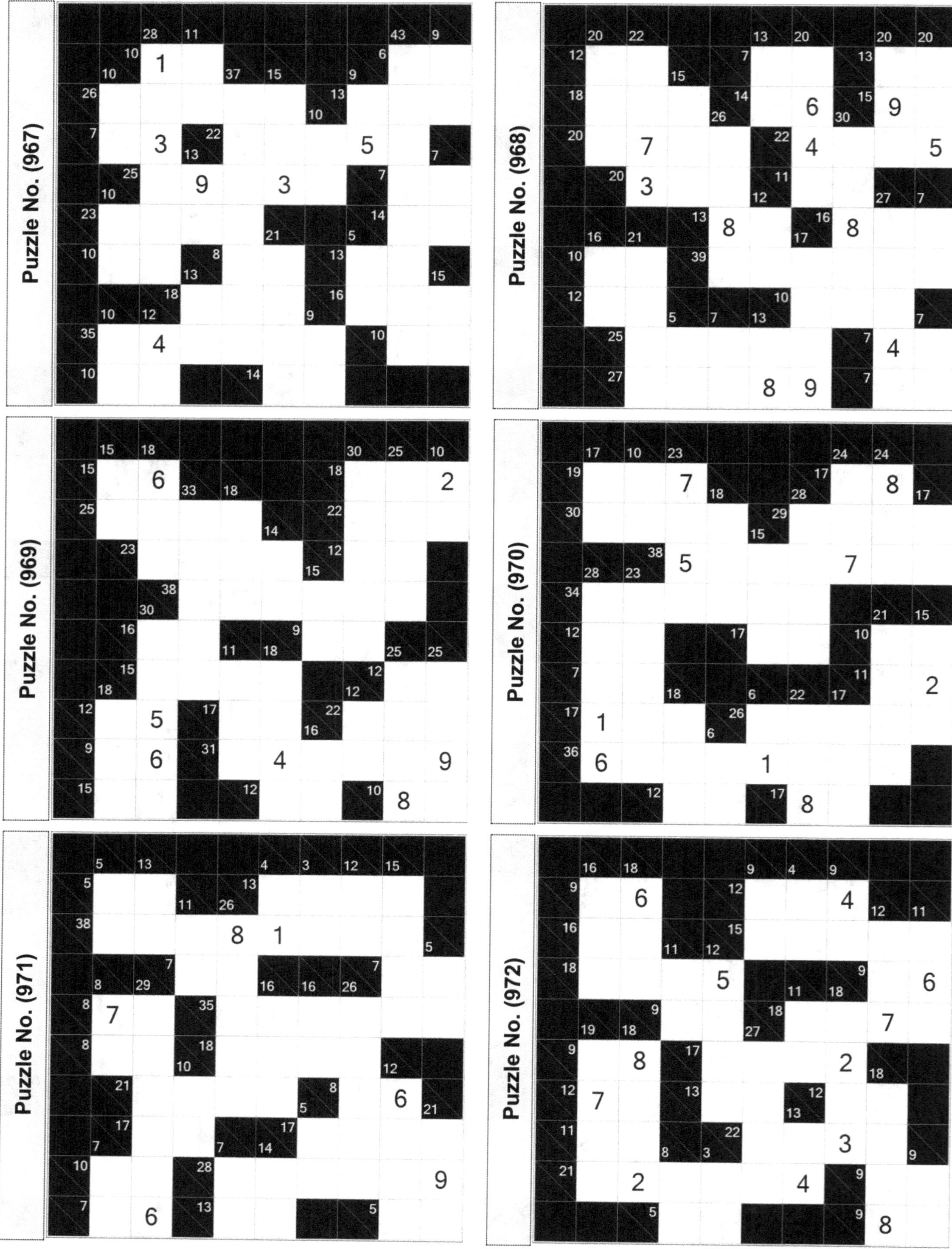

Puzzle No. (967)

Puzzle No. (968)

Puzzle No. (969)

Puzzle No. (970)

Puzzle No. (971)

Puzzle No. (972)

Solution on Page (218)

(164)

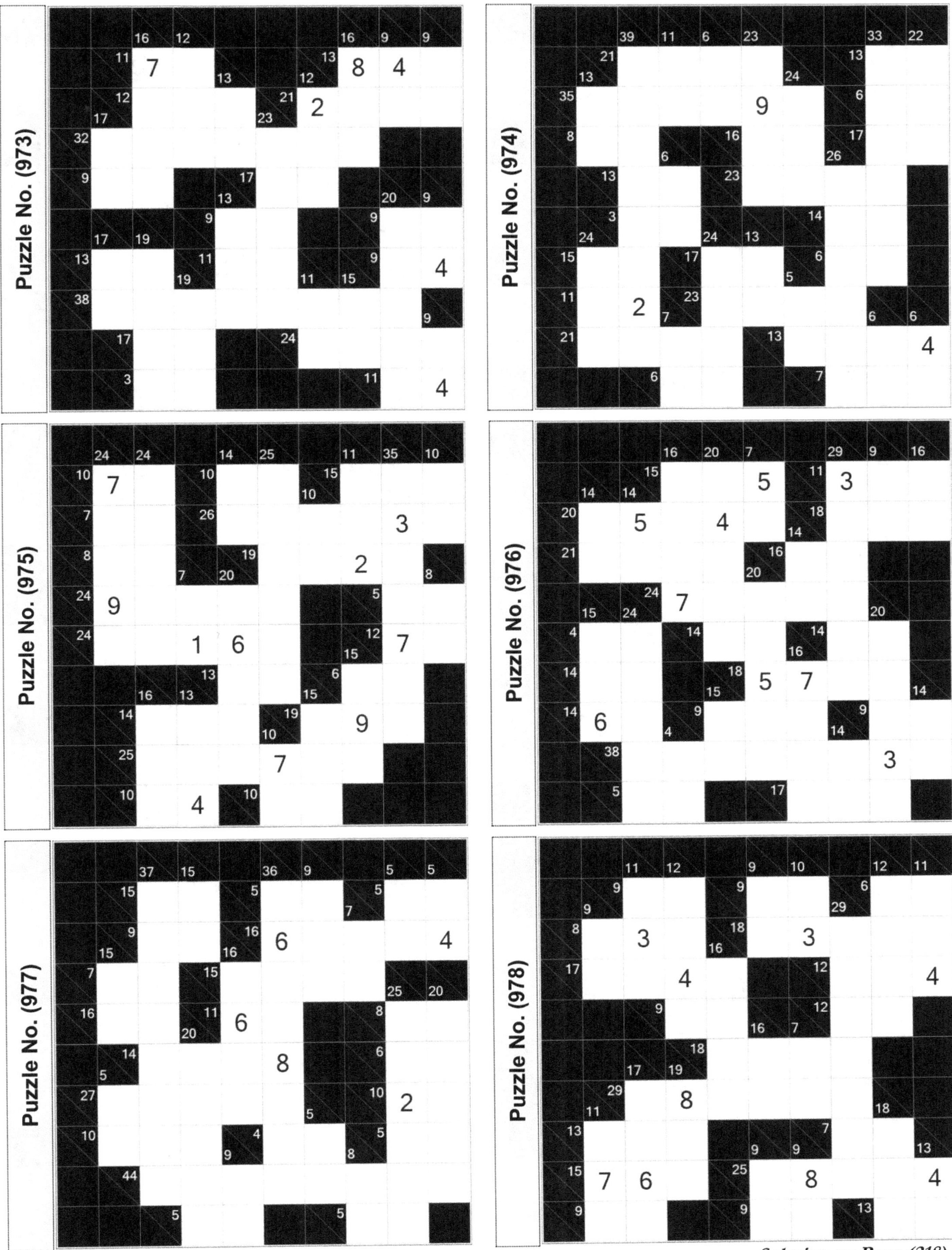

Puzzle No. (973)

Puzzle No. (974)

Puzzle No. (975)

Puzzle No. (976)

Puzzle No. (977)

Puzzle No. (978)

Solution on Page (218)

(165)

Puzzle No. (979)

Puzzle No. (980)

Puzzle No. (981)

Puzzle No. (982)

Puzzle No. (983)

Puzzle No. (984)

Solution on Pages (218-219)

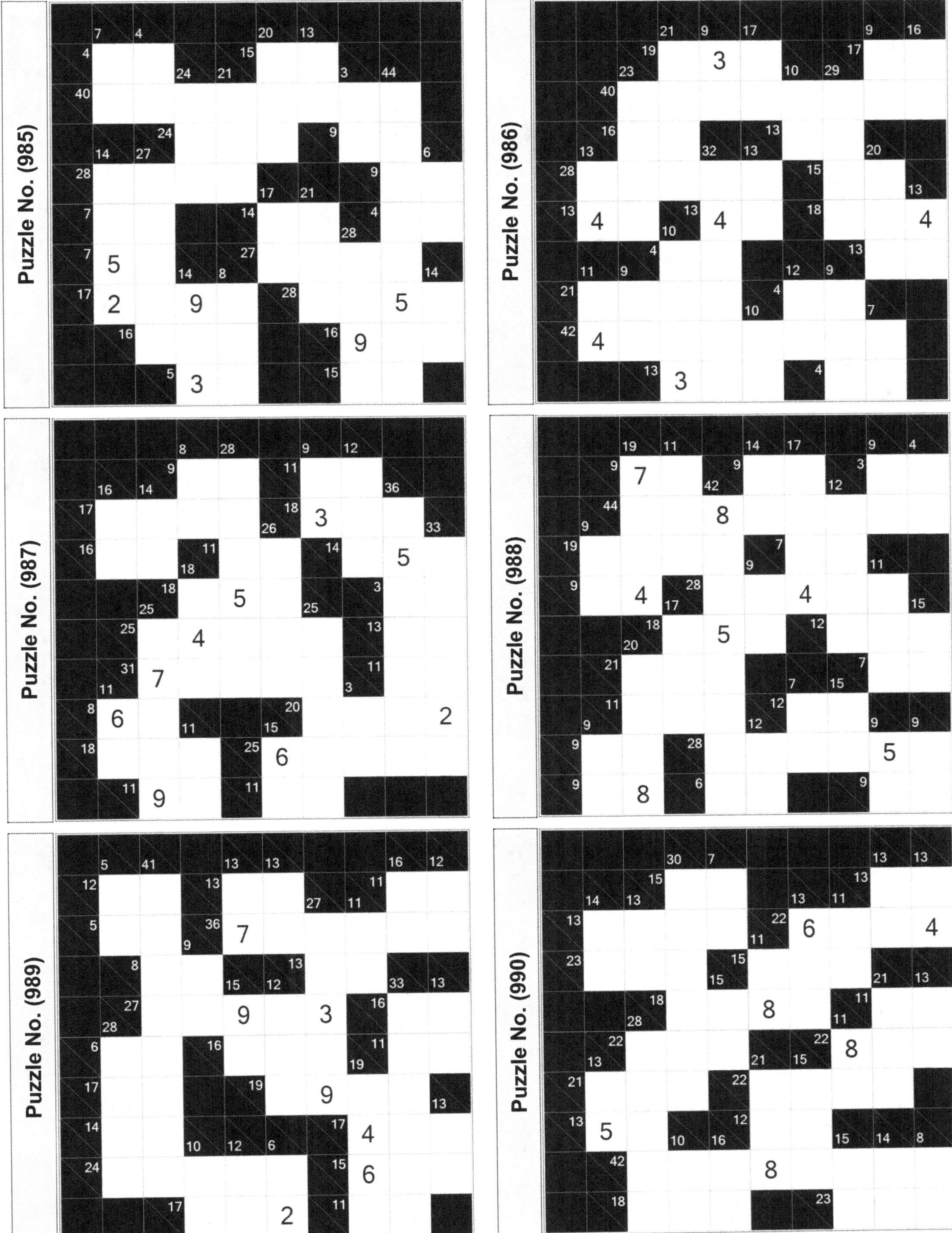

Puzzle No. (985)
Puzzle No. (986)
Puzzle No. (987)
Puzzle No. (988)
Puzzle No. (989)
Puzzle No. (990)
Solution on Page (219)

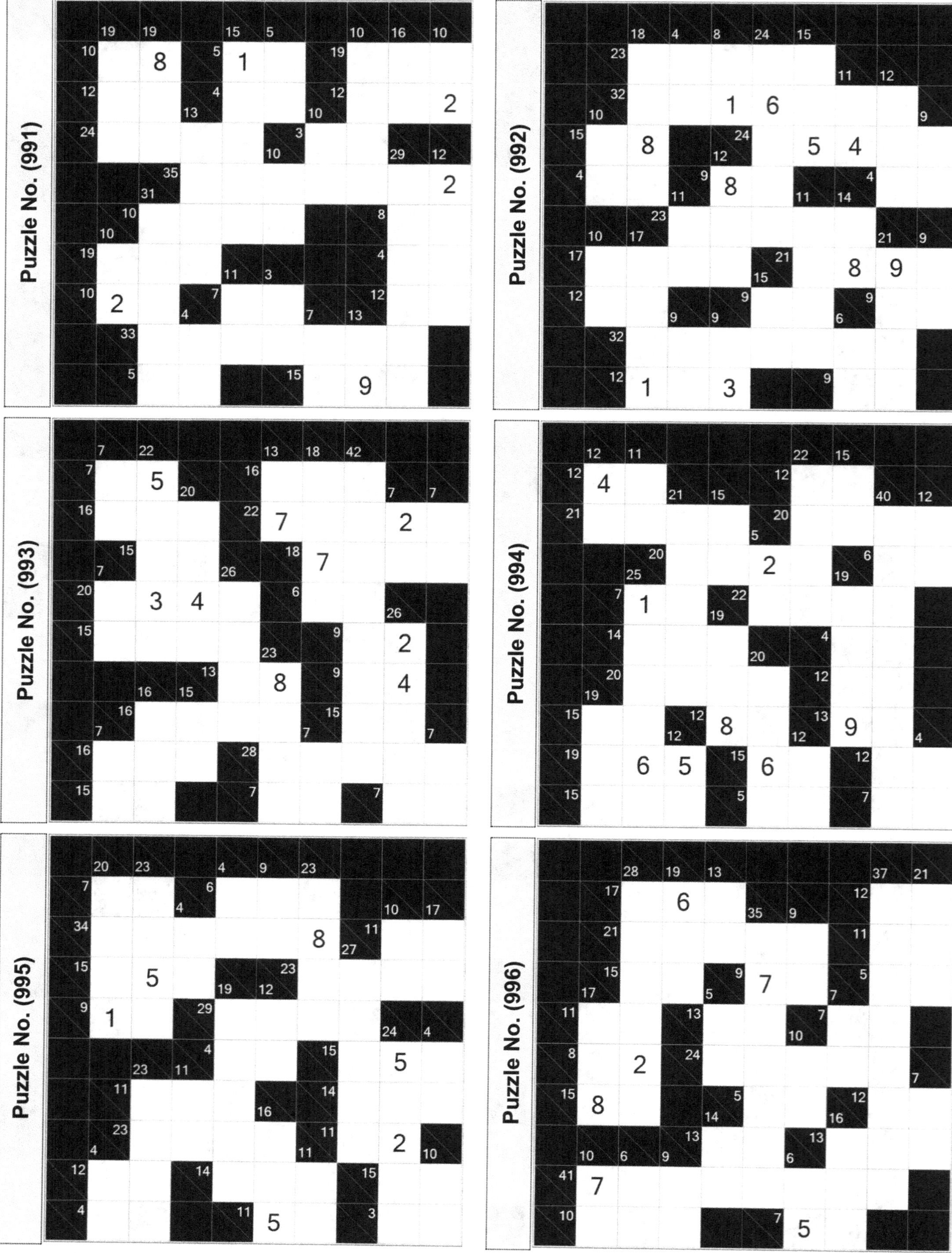

Solution on Page (219)

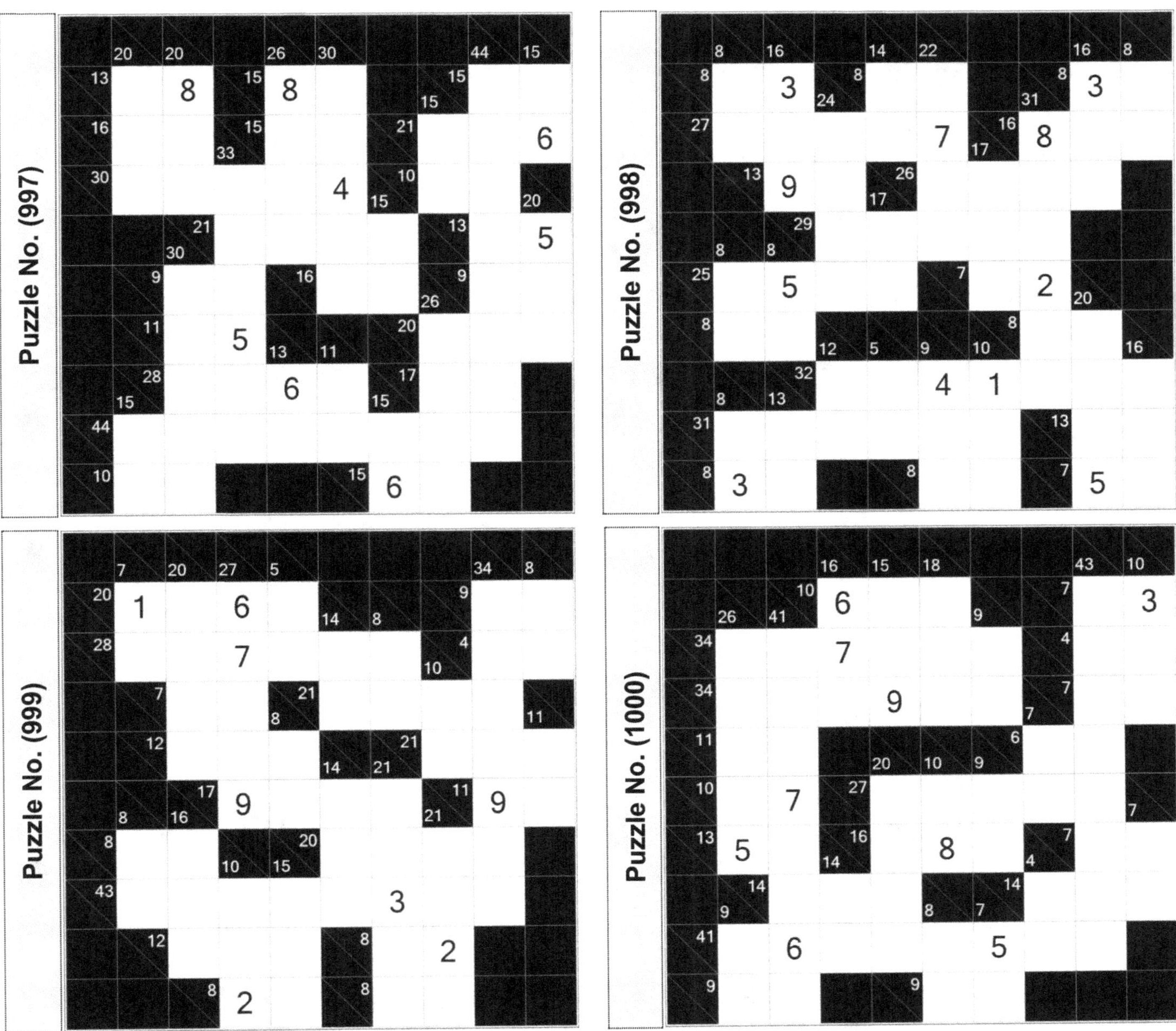

Puzzle No. (997)

Puzzle No. (998)

Puzzle No. (999)

Puzzle No. (1000)

Solution on Page (219)

::::: *Puzzle (21)* ::::: ::::: *Puzzle (22)* ::::: ::::: *Puzzle (23)* ::::: ::::: *Puzzle (24)* :::::

::::: *Puzzle (25)* ::::: ::::: *Puzzle (26)* ::::: ::::: *Puzzle (27)* ::::: ::::: *Puzzle (28)* :::::

::::: *Puzzle (29)* ::::: ::::: *Puzzle (30)* ::::: ::::: *Puzzle (31)* ::::: ::::: *Puzzle (32)* :::::

::::: *Puzzle (33)* ::::: ::::: *Puzzle (34)* ::::: ::::: *Puzzle (35)* ::::: ::::: *Puzzle (36)* :::::

::::: *Puzzle (37)* ::::: ::::: *Puzzle (38)* ::::: ::::: *Puzzle (39)* ::::: ::::: *Puzzle (40)* :::::

::::: Puzzle (61) :::::
::::: Puzzle (62) :::::
::::: Puzzle (63) :::::
::::: Puzzle (64) :::::
::::: Puzzle (65) :::::
::::: Puzzle (66) :::::
::::: Puzzle (67) :::::
::::: Puzzle (68) :::::
::::: Puzzle (69) :::::
::::: Puzzle (70) :::::
::::: Puzzle (71) :::::
::::: Puzzle (72) :::::
::::: Puzzle (73) :::::
::::: Puzzle (74) :::::
::::: Puzzle (75) :::::
::::: Puzzle (76) :::::
::::: Puzzle (77) :::::
::::: Puzzle (78) :::::
::::: Puzzle (79) :::::
::::: Puzzle (80) :::::

::::: *Puzzle (81)* ::::: ::::: *Puzzle (82)* ::::: ::::: *Puzzle (83)* ::::: ::::: *Puzzle (84)* :::::

::::: *Puzzle (85)* ::::: ::::: *Puzzle (86)* ::::: ::::: *Puzzle (87)* ::::: ::::: *Puzzle (88)* :::::

::::: *Puzzle (89)* ::::: ::::: *Puzzle (90)* ::::: ::::: *Puzzle (91)* ::::: ::::: *Puzzle (92)* :::::

::::: *Puzzle (93)* ::::: ::::: *Puzzle (94)* ::::: ::::: *Puzzle (95)* ::::: ::::: *Puzzle (96)* :::::

::::: *Puzzle (97)* ::::: ::::: *Puzzle (98)* ::::: ::::: *Puzzle (99)* ::::: ::::: *Puzzle (100)* :::::

::::: Puzzle (121) :::::
::::: Puzzle (122) :::::
::::: Puzzle (123) :::::
::::: Puzzle (124) :::::
::::: Puzzle (125) :::::
::::: Puzzle (126) :::::
::::: Puzzle (127) :::::
::::: Puzzle (128) :::::
::::: Puzzle (129) :::::
::::: Puzzle (130) :::::
::::: Puzzle (131) :::::
::::: Puzzle (132) :::::
::::: Puzzle (133) :::::
::::: Puzzle (134) :::::
::::: Puzzle (135) :::::
::::: Puzzle (136) :::::
::::: Puzzle (137) :::::
::::: Puzzle (138) :::::
::::: Puzzle (139) :::::
::::: Puzzle (140) :::::

(176)

(182)

::::: Puzzle (261) :::::
::::: Puzzle (262) :::::
::::: Puzzle (263) :::::
::::: Puzzle (264) :::::
::::: Puzzle (265) :::::
::::: Puzzle (266) :::::
::::: Puzzle (267) :::::
::::: Puzzle (268) :::::
::::: Puzzle (269) :::::
::::: Puzzle (270) :::::
::::: Puzzle (271) :::::
::::: Puzzle (272) :::::
::::: Puzzle (273) :::::
::::: Puzzle (274) :::::
::::: Puzzle (275) :::::
::::: Puzzle (276) :::::
::::: Puzzle (277) :::::
::::: Puzzle (278) :::::
::::: Puzzle (279) :::::
::::: Puzzle (280) :::::

(186)

::::: *Puzzle (341)* ::::: ::::: *Puzzle (342)* ::::: ::::: *Puzzle (343)* ::::: ::::: *Puzzle (344)* :::::

::::: *Puzzle (345)* ::::: ::::: *Puzzle (346)* ::::: ::::: *Puzzle (347)* ::::: ::::: *Puzzle (348)* :::::

::::: *Puzzle (349)* ::::: ::::: *Puzzle (350)* ::::: ::::: *Puzzle (351)* ::::: ::::: *Puzzle (352)* :::::

::::: *Puzzle (353)* ::::: ::::: *Puzzle (354)* ::::: ::::: *Puzzle (355)* ::::: ::::: *Puzzle (356)* :::::

::::: *Puzzle (357)* ::::: ::::: *Puzzle (358)* ::::: ::::: *Puzzle (359)* ::::: ::::: *Puzzle (360)* :::::

(187)

::::: Puzzle (421) :::::
::::: Puzzle (422) :::::
::::: Puzzle (423) :::::
::::: Puzzle (424) :::::
::::: Puzzle (425) :::::
::::: Puzzle (426) :::::
::::: Puzzle (427) :::::
::::: Puzzle (428) :::::
::::: Puzzle (429) :::::
::::: Puzzle (430) :::::
::::: Puzzle (431) :::::
::::: Puzzle (432) :::::
::::: Puzzle (433) :::::
::::: Puzzle (434) :::::
::::: Puzzle (435) :::::
::::: Puzzle (436) :::::
::::: Puzzle (437) :::::
::::: Puzzle (438) :::::
::::: Puzzle (439) :::::
::::: Puzzle (440) :::::

::::: Puzzle (481) ::::: ::::: Puzzle (482) ::::: ::::: Puzzle (483) ::::: ::::: Puzzle (484) :::::
::::: Puzzle (485) ::::: ::::: Puzzle (486) ::::: ::::: Puzzle (487) ::::: ::::: Puzzle (488) :::::
::::: Puzzle (489) ::::: ::::: Puzzle (490) ::::: ::::: Puzzle (491) ::::: ::::: Puzzle (492) :::::
::::: Puzzle (493) ::::: ::::: Puzzle (494) ::::: ::::: Puzzle (495) ::::: ::::: Puzzle (496) :::::
::::: Puzzle (497) ::::: ::::: Puzzle (498) ::::: ::::: Puzzle (499) ::::: ::::: Puzzle (500) :::::

(194)

::::: Puzzle (541) :::::
::::: Puzzle (542) :::::
::::: Puzzle (543) :::::
::::: Puzzle (544) :::::
::::: Puzzle (545) :::::
::::: Puzzle (546) :::::
::::: Puzzle (547) :::::
::::: Puzzle (548) :::::
::::: Puzzle (549) :::::
::::: Puzzle (550) :::::
::::: Puzzle (551) :::::
::::: Puzzle (552) :::::
::::: Puzzle (553) :::::
::::: Puzzle (554) :::::
::::: Puzzle (555) :::::
::::: Puzzle (556) :::::
::::: Puzzle (557) :::::
::::: Puzzle (558) :::::
::::: Puzzle (559) :::::
::::: Puzzle (560) :::::

Puzzle (561)
Puzzle (562)
Puzzle (563)
Puzzle (564)
Puzzle (565)
Puzzle (566)
Puzzle (567)
Puzzle (568)
Puzzle (569)
Puzzle (570)
Puzzle (571)
Puzzle (572)
Puzzle (573)
Puzzle (574)
Puzzle (575)
Puzzle (576)
Puzzle (577)
Puzzle (578)
Puzzle (579)
Puzzle (580)

::::: Puzzle (601) :::::
::::: Puzzle (602) :::::
::::: Puzzle (603) :::::
::::: Puzzle (604) :::::
::::: Puzzle (605) :::::
::::: Puzzle (606) :::::
::::: Puzzle (607) :::::
::::: Puzzle (608) :::::
::::: Puzzle (609) :::::
::::: Puzzle (610) :::::
::::: Puzzle (611) :::::
::::: Puzzle (612) :::::
::::: Puzzle (613) :::::
::::: Puzzle (614) :::::
::::: Puzzle (615) :::::
::::: Puzzle (616) :::::
::::: Puzzle (617) :::::
::::: Puzzle (618) :::::
::::: Puzzle (619) :::::
::::: Puzzle (620) :::::

::::: Puzzle (621) :::::
::::: Puzzle (622) :::::
::::: Puzzle (623) :::::
::::: Puzzle (624) :::::
::::: Puzzle (625) :::::
::::: Puzzle (626) :::::
::::: Puzzle (627) :::::
::::: Puzzle (628) :::::
::::: Puzzle (629) :::::
::::: Puzzle (630) :::::
::::: Puzzle (631) :::::
::::: Puzzle (632) :::::
::::: Puzzle (633) :::::
::::: Puzzle (634) :::::
::::: Puzzle (635) :::::
::::: Puzzle (636) :::::
::::: Puzzle (637) :::::
::::: Puzzle (638) :::::
::::: Puzzle (639) :::::
::::: Puzzle (640) :::::

(201)

::::: *Puzzle (641)* ::::: ::::: *Puzzle (642)* ::::: ::::: *Puzzle (643)* ::::: ::::: *Puzzle (644)* :::::

::::: *Puzzle (645)* ::::: ::::: *Puzzle (646)* ::::: ::::: *Puzzle (647)* ::::: ::::: *Puzzle (648)* :::::

::::: *Puzzle (649)* ::::: ::::: *Puzzle (650)* ::::: ::::: *Puzzle (651)* ::::: ::::: *Puzzle (652)* :::::

::::: *Puzzle (653)* ::::: ::::: *Puzzle (654)* ::::: ::::: *Puzzle (655)* ::::: ::::: *Puzzle (656)* :::::

::::: *Puzzle (657)* ::::: ::::: *Puzzle (658)* ::::: ::::: *Puzzle (659)* ::::: ::::: *Puzzle (660)* :::::

::::: Puzzle (681) :::::
::::: Puzzle (682) :::::
::::: Puzzle (683) :::::
::::: Puzzle (684) :::::
::::: Puzzle (685) :::::
::::: Puzzle (686) :::::
::::: Puzzle (687) :::::
::::: Puzzle (688) :::::
::::: Puzzle (689) :::::
::::: Puzzle (690) :::::
::::: Puzzle (691) :::::
::::: Puzzle (692) :::::
::::: Puzzle (693) :::::
::::: Puzzle (694) :::::
::::: Puzzle (695) :::::
::::: Puzzle (696) :::::
::::: Puzzle (697) :::::
::::: Puzzle (698) :::::
::::: Puzzle (699) :::::
::::: Puzzle (700) :::::

::::: Puzzle (741) :::::
::::: Puzzle (742) :::::
::::: Puzzle (743) :::::
::::: Puzzle (744) :::::
::::: Puzzle (745) :::::
::::: Puzzle (746) :::::
::::: Puzzle (747) :::::
::::: Puzzle (748) :::::
::::: Puzzle (749) :::::
::::: Puzzle (750) :::::
::::: Puzzle (751) :::::
::::: Puzzle (752) :::::
::::: Puzzle (753) :::::
::::: Puzzle (754) :::::
::::: Puzzle (755) :::::
::::: Puzzle (756) :::::
::::: Puzzle (757) :::::
::::: Puzzle (758) :::::
::::: Puzzle (759) :::::
::::: Puzzle (760) :::::

(207)

::::: Puzzle (761) :::::
::::: Puzzle (762) :::::
::::: Puzzle (763) :::::
::::: Puzzle (764) :::::
::::: Puzzle (765) :::::
::::: Puzzle (766) :::::
::::: Puzzle (767) :::::
::::: Puzzle (768) :::::
::::: Puzzle (769) :::::
::::: Puzzle (770) :::::
::::: Puzzle (771) :::::
::::: Puzzle (772) :::::
::::: Puzzle (773) :::::
::::: Puzzle (774) :::::
::::: Puzzle (775) :::::
::::: Puzzle (776) :::::
::::: Puzzle (777) :::::
::::: Puzzle (778) :::::
::::: Puzzle (779) :::::
::::: Puzzle (780) :::::

::::: Puzzle (881) :::::
::::: Puzzle (882) :::::
::::: Puzzle (883) :::::
::::: Puzzle (884) :::::
::::: Puzzle (885) :::::
::::: Puzzle (886) :::::
::::: Puzzle (887) :::::
::::: Puzzle (888) :::::
::::: Puzzle (889) :::::
::::: Puzzle (890) :::::
::::: Puzzle (891) :::::
::::: Puzzle (892) :::::
::::: Puzzle (893) :::::
::::: Puzzle (894) :::::
::::: Puzzle (895) :::::
::::: Puzzle (896) :::::
::::: Puzzle (897) :::::
::::: Puzzle (898) :::::
::::: Puzzle (899) :::::
::::: Puzzle (900) :::::

::::: Puzzle (981) :::::
::::: Puzzle (982) :::::
::::: Puzzle (983) :::::
::::: Puzzle (984) :::::
::::: Puzzle (985) :::::
::::: Puzzle (986) :::::
::::: Puzzle (987) :::::
::::: Puzzle (988) :::::
::::: Puzzle (989) :::::
::::: Puzzle (990) :::::
::::: Puzzle (991) :::::
::::: Puzzle (992) :::::
::::: Puzzle (993) :::::
::::: Puzzle (994) :::::
::::: Puzzle (995) :::::
::::: Puzzle (996) :::::
::::: Puzzle (997) :::::
::::: Puzzle (998) :::::
::::: Puzzle (999) :::::
::::: Puzzle (1000) :::::

Table (1-A) : Kakuro Combinations

Sum	No. of Squares			
	2	3	4	5
3	12			
4	13			
5	14, 23			
6	15, 24	123		
7	16, 25, 34	124		
8	17, 26, 35	125, 134		
9	18, 27, 36, 45	126, 135, 234		
10	19, 28, 37, 46	127, 136, 145, 235	1234	
11	29, 38, 47, 56	128, 137, 146, 236, 245	1235	
12	39, 48, 57	129, 138, 147, 156, 237, 246, 345	1236, 1245	
13	49, 58, 67	139, 148, 157, 238, 247, 256, 346	1237, 1246, 1345	
14	59, 68	149, 158, 167, 239, 248, 257, 347, 356	1238, 1247, 1256, 1346, 2345	
15	69, 78	159, 168, 249, 258, 267, 348, 357, 456	1239, 1248, 1257, 1347, 1356, 2346	12345
16	79	169, 178, 259, 268, 349, 358, 367, 457	1249, 1258, 1267, 1348, 1357, 1456, 2347, 2356	12346
17	89	179, 269, 278, 359, 368, 458, 467	1259, 1268, 1349, 1358, 1367, 1457, 2348, 2357, 2456	12347, 12356
18		189, 279, 369, 378, 459, 468, 567	1269, 1278, 1359, 1368, 1458, 1467, 2349, 2358, 2367, 2457, 3456	12348, 12357, 12456
19		289, 379, 469, 478, 568	1279, 1369, 1378, 1459, 1468, 1567, 2359, 2368, 2458, 2467, 3457	12349, 12358, 12367, 12457, 13456
20		389, 479, 569, 578	1289, 1379, 1469, 1478, 1568, 2369, 2378, 2459, 2468, 2567, 3458, 3467	12359, 12368, 12458, 12467, 13457, 23456

Table (1-B) : Kakuro Combinations

Sum	No. of Squares				
	3	4	5	6	7
21	489, 579, 678	1389, 1479, 1569, 1578, 2379, 2469, 2478, 2568, 3459, 3468, 3567	12369, 12378, 12459, 12468, 12567, 13458, 13467, 23457	123456	
22	589, 679	1489, 1579, 1678, 2389, 2479, 2569, 2578, 3469, 3478, 3568, 4567	12379, 12469, 12478, 12568, 13459, 13468, 13567, 23458, 23467	123457	
23	689	1589, 1679, 2489, 2579, 2678, 3479, 3569, 3578, 4568	12389, 12479, 12569, 12578, 13469, 13478, 13568, 14567, 23459, 23468, 23567	123458, 123467	
24	789	1689, 2589, 2679, 3489, 3579, 3678, 4569, 4578	12489, 12579, 12678, 13479, 13569, 13578, 14568, 23469, 23478, 23568, 24567	123459, 123468, 123567	
25		1789, 2689, 3589, 3679, 4579, 4678	12589, 12679, 13489, 13579, 13678, 14569, 14578, 23479, 23569, 23578, 24568, 34567	123469, 123478, 123568, 124567	
26		2789, 3689, 4589, 4679, 5678	12689, 13589, 13679, 14579, 14678, 23489, 23579, 23678, 24569, 24578, 34568	123479, 123569, 123578, 124568, 134567	
27		3789, 4689, 5679	12789, 13689, 14589, 14679, 15678, 23589, 23679, 24579, 24678, 34569, 34578	123489, 123579, 123678, 124569, 124578, 134568, 234567	
28		4789, 5689	13789, 14689, 15679, 23689, 24589, 24679, 25678, 34579, 34678	123589, 123679, 124579, 124678, 134569, 134578, 234568	1234567
29		5789	14789, 15689, 23789, 24689, 25679, 34589, 34679, 35678	123689, 124589, 124679, 125678, 134579, 134678, 234569, 234578	1234568
30		6789	15789, 24789, 25689, 34689, 35679, 45678	123789, 124689, 125679, 134589, 134679, 135678, 234579, 234678	1234569, 1234578

Table (1-C) : Kakuro Combinations

Sum	No. of Squares				
	5	6	7	8	9
31	16789, 25789, 34789, 35689, 45679	124789, 125689, 134689, 135679, 145678, 234589, 234679, 235678	1234579, 1234678		
32	26789, 35789, 45689	125789, 134789, 135689, 145679, 234689, 235679, 245678	1234589, 1234679, 1235678		
33	36789, 45789	126789, 135789, 145689, 234789, 235689, 245679, 345678	1234689, 1235679, 1245678		
34	46789	136789, 145789, 235789, 245689, 345679	1234789, 1235689, 1245679, 1345678		
35	56789	146789, 236789, 245789, 345689	1235789, 1245689, 1345679, 2345678		
36		156789, 246789, 345789	1236789, 1245789, 1345689, 2345679	12345678	
37		256789, 346789	1246789, 1345789, 2345689	12345679	
38		356789	1256789, 1346789, 2345789	12345689	
39		456789	1356789, 2346789	12345789	
40			1456789, 2356789	12346789	
41			2456789	12356789	
42			3456789	12456789	
43				13456789	
44				23456789	
45					123456789

Table (2) : Kakuro Magic Squares*

Sum	No. of Squares	Combination	Sum	No. of Squares	Combination
3	2	12	22	6	123457
4	2	13	38	6	356789
16	2	79	39	6	456789
17	2	89	28	7	1234567
6	3	123	29	7	1234568
7	3	124	41	7	2456789
23	3	689	42	7	3456789
24	3	789	36	8	12345678
10	4	1234	37	8	12345679
11	4	1235	38	8	12345689
29	4	5789	39	8	12345789
30	4	6789	40	8	12346789
15	5	12345	41	8	12356789
16	5	12346	42	8	12456789
34	5	46789	43	8	13456789
35	5	56789	44	8	23456789
21	6	123456	45	9	123456789

* This table shows special situations where only a single combination of numbers can fit into a square of a given length.